Bioenergetics at a Glance

D1333586

Bioenergetics at a Glance

DAVID A. HARRIS MA, DPhil
Department of Biochemistry
University of Oxford

b
**Blackwell
Science**

© 1995 by Blackwell Science Ltd
Editorial Offices:
Osney Mead, Oxford OX2 0EL
25 John Street, London WC1N 2BL
23 Ainslie Place, Edinburgh EH3 6AJ
238 Main Street, Cambridge, Massachusetts 02142, USA
54 University Street, Carlton, Victoria 3053, Australia

Other Editorial Offices:
Arnette Blackwell SA
  1, rue de Lille, 75007 Paris, France

Blackwell Wissenschafts-Verlag GmbH
  Kurfürstendamm 57, 10707 Berlin, Germany

Zehetnergasse 6, A-1140 Wien, Austria

First published 1995
Reprinted 1996
Four Dragons edition 1995

Set by DP Photosetting, Aylesbury, Bucks
Printed and bound in Great Britain
at the University Press, Cambridge

DISTRIBUTORS

Marston Book Services Ltd
PO Box 87, Oxford OX2 0DT
(*Orders*: Tel: 01865 791155
      Fax: 01865 791927
      Telex: 837515)

North America
  Blackwell Science, Inc.
  238 Main Street, Cambridge, MA 02142
  (*Orders*: Tel: 800 215-1000
        617 876-7000
      Fax: 617 492-5263)

Australia
  Blackwell Science Pty Ltd
  54 University Street, Carlton, Victoria 3053
  (*Orders*: Tel: 03 9347-0300
      Fax: 03 9349-3016)

A catalogue record for this title is available from the British Library
and the Library of Congress

ISBN 0-632-02388-0 (Blackwell Science)
    0-632-03423-8 (Four Dragons)

Contents

Preface

Bioenergetics may be defined as the study of energy transformations within living systems. This is an attractive definition for the bioenergeticist, as it encompasses virtually the whole of biochemistry and biology; all life processes represent a struggle against the Second Law, using exergonic reactions as a source of energy. Harold has designated bioenergetics the 'scientific study of vitality', in contrast to molecular biology, the 'study of biological structures'.

Unfortunately, this definition is unattractive for an author. Bioenergetics textbooks cannot be infinitely long, nor can they turn into biochemistry textbooks—there are plenty of those already. Instead, the definition is conventionally restricted to processes generating ATP, with processes using ATP (all the rest of biochemistry) being omitted. University courses in bioenergetics are, typically, further restricted to considering ATP synthesis linked to redox reactions, such as the oxidation of NADH. This area, therefore, forms the central core of the present text (Sections 10–30), and can be read in isolation—with only the occasional need for clarification from sections before and after.

This restriction is, however, too severe. The role of ion gradients, championed by Peter Mitchell, requires there to be a close interrelationship between such gradients and both redox reactions and ATP hydrolysis. Thus there are, included in the present text, analogous systems—generation of ion gradients by ATP hydrolysis or by light—and other possible uses of ion gradients—in transport, heat generation or motion. Similarly, present day bioenergetics is the product of a long evolutionary history, and, in particular, the remarkable endosymbiotic origin of mitochondria and chloroplasts; this too is covered. I have reluctantly, but conventionally drawn the line above ATP-driven motility systems such as muscle and eukaryotic flagellae.

This is a good time to write a textbook on bioenergetics. Advances in molecular biology over the past 10 years have allowed us to move away from 'black boxes' and 'squiggles' (~) and give our ideas of mechanism a structural basis—still oversimplified in many cases, but at least lending an air of clarity. I have thus attempted, wherever possible, to combine ideas of structure and mechanism—to provide the framework without losing track of the vitality.

Bioenergetics at a Glance is structured as a modular text, with topics discussed in one—or occasionally more—double page spreads or modules. Apart from a certain arbitrariness in the choice of boundaries, the major drawback to this approach is the inability to explore controversies in any depth. In such cases, I regret any oversimplifications, or overextrapolation on my part, which may have resulted. I justify them in the aim of clarifying the underlying principles.

The advantages of this structure lie in its ease of use. Topics for study, and for revision, are easily identified and can be studied with little reference to the rest of the text (except perhaps as a glossary). I have attempted to grade central topics that span several modules; the first one or two modules are relatively simpler, and can be read alone if only an introduction to the topic is needed. In addition, space restrictions ensure that key points are emphasized and illustrative examples provided; for an exhaustive coverage, additional references are provided.

I have also included (in Sections 2–6) an outline of quantitative solution thermodynamics. Again, this section may be omitted from an initial reading if only qualitative bioenergetics is to be studied. However, I believe that, without understanding the relevant thermodynamics, the full elegance of nature's coupling systems cannot be appreciated. The key word here is relevant—these sections are structured in terms of potential energy and chemical potentials, and require no knowledge of heat engines, reversible cycles, and the other paraphernalia so beloved of the chemist. Living systems cannot vary temperature or pressure at will, and so their thermodynamics can be relatively simple, as I have attempted to demonstrate.

David Harris
Oxford

1 Life, energy and metabolism

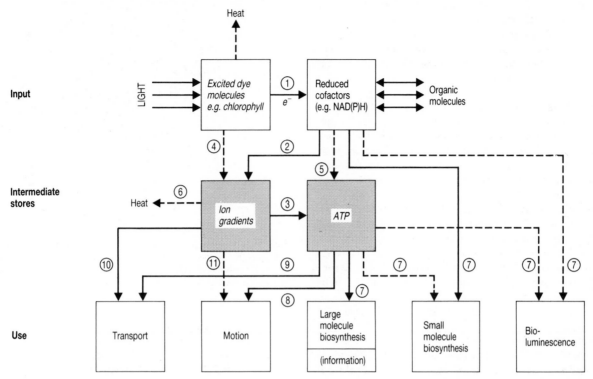

Fig. 1.1. Energy flow in biological systems.
① Photosynthetic electron transfer (Sections 15–18).
② Respiratory chains (Sections 11–14).
③ Transduction of H^+ gradient into ATP (Sections 27–28).
④ Bacteriorhodopsin H^+ pump (Section 40).
⑤ Substrate level phosphorylation (Section 8).
⑥ Heat production in brown adipose tissue (Section 37).
⑦ Metabolic pathways (Section 1).
⑧ Muscle contraction.
⑨ ATP driven pumps (Sections 43–44).
⑩ Gradient driven porters (Section 38).
⑪ Bacterial flagellar motor (Section 39).

Primary energy sources

Life is an uphill struggle

Living organisms require a continual supply of energy to build up complex molecules, to organize them, to move around, to maintain body temperature etc.—in short, to construct local areas of *specialization* in a large chaotic universe. This energy must be derived from the world outside the organism. Indeed, it is a characteristic of life that it can harvest energy from the environment for these purposes. It is the mechanisms involved in *energy trapping* and *interconversion* that define the area of this text, **bioenergetics**; how this energy is utilized within organisms constitutes the remainder of biochemistry.

There are two sources of energy that can be tapped by living organisms—light energy (from the sun), and energy from oxidations (Fig. 1.1).

Light absorbtion

Photons provide a potential source of energy for all living organisms. However, only photosynthetic organisms can use this energy. To use it they require (i) a *dye* (usually chlorophyll) to absorb the light, and (ii) a *system to convert this energy into a useful form*.

Most dyes will dissipate absorbed energy, as heat. While this is occasionally useful to organisms (for example, in raising the temperature of basking reptiles), it is generally more convenient to *trap* the absorbed energy as chemical changes. Since absorbtion of light involves exciting an *electron* from one orbital to another of higher energy, energy is commonly trapped by using this excited electron as a **reductant**. The reduced compound then yields its energy in subsequent **oxidations**.

The utility of this energy trapping process lies in its **time scale**. Excited chlorophyll, for example, has a life time of about 10^{-10} s, and would be difficult to use as a source of energy in biochemical processes. (Enzyme turnover, for example, takes about 10^{-2} s). Reduced organic molecules, like fats, in contrast survive for several months or years (10^8 s), and can be used when required.

Oxidation reactions

We live in a highly oxidizing environment; 23 per cent of our atmosphere is oxygen. This means that the oxidation of most organic compounds (e.g. glucose, fats) is thermodynamically favourable, and can serve as a source of energy. Even when no oxygen is available, other oxidized elements (SO_4^{2-}, NO_3^-) may accept electrons and oxidize organic compounds to yield energy (Section 14).

Animals, and most bacteria, obtain all their energy by importing organic compounds (food) and oxidizing them in this way. Plants obtain their energy by absorbing light—but, as noted above, this energy is used primarily to generate reduced organic molecules. It is the oxidation of these compounds, again, that yields the energy necessary for driving the plant's biochemistry.

Energy consuming processes (see Fig. 1.1)

Production of small organic molecules

From organic compounds like glucose (or even from CO_2), living organisms can create a wide variety of molecules—long chain fatty acids, linear and heterocyclic C–N compounds (e.g. pyrimidines), multiple ring systems (e.g. steroids) etc. All of these are thermodynamically *less stable*, in an oxidizing environment, than the starting material, and hence their formation is an uphill process.

These reactions are driven, largely, by the participation of a powerful reducing agent, **NADPH**, which is maintained in the cell far away from its equilibrium with oxygen. Pathways for the **biosynthesis** of small molecules invariably involve reduction by NADPH and a movement away from the oxidized form of carbon, CO_2. A further participant in most of these pathways is the **acid anhydride**, ATP (see Section 7).

Production of macromolecules

From small organic molecules (monosaccharides, amino acids, etc.), macromolecules (polysaccharides, proteins, etc.) are built up by *condensation reactions*, in which water is removed. Here, the major energy source is ATP; as an acid anhydride it can act as a **dehydrating agent** even (since it is not in equilibrium with water) in an aqueous environment.

In the case of proteins and nucleic acids, it is not sufficient to join any pair of amino acids or nucleotides; we need to condense the *correct* ones. This entails mechanisms of *selection* and *proofreading* to ensure the correct reaction. In these processes, energy is converted directly into **information**. This energy is also supplied by ATP (or in the case of protein synthesis, by its close analogue, GTP), but in this case is used to drive steric matching between the reacting species and the conformation of a specific protein (transducer) molecule.

Transport of ions and molecules

Cells, or regions of cells, typically accumulate the molecules and ions that they need (e.g. glucose, amino acids, Ca^{2+}), and expel others (water, Na^+). In doing this, they create **gradients** of metabolites and ions across their plasma membranes and internal membranes. They may also use such gradients for **information transfer** ('signalling').

To produce and maintain such gradients, the cell must move, or *pump* molecules from regions of low concentration to regions of higher concentration, against the tendency of the molecules to equilibrate. This is clearly a process that requires energy.

Movement of cells and organisms

Muscle contraction, **swimming** by sperm or bacteria, and the **beating of cilia** on cell surfaces are all processes which require a conversion of metabolic energy into obvious, mechanical, work. ATP, via protein conformational changes, is commonly the energy source.

Light emission

Photosynthesis traps photons, and conserves their energy by reduction of organic molecules. A few organisms (fireflies, jellyfish, some bacteria) are able to *reverse* this process and **generate light**. The energy, in this case, is derived from oxidation, by O_2, of a complex organic molecule, known as a *luciferin*. The precise chemical nature of the luciferin varies with species.

Patterns in metabolism

Energy partitioning

Energy required for life processes (above) is provided by the oxidative breakdown of organic molecules—**catabolism**. However, the chemistry of catabolism is complex; glucose, for example, is oxidized to CO_2 in a series of 22 different reactions. Why has such complexity evolved?

There is probably no one factor that explains the existing complexity of catabolism; the *chemical nature of enzymes* themselves, and *evolutionary mechanisms*, will both restrict the reactions available to be selected. However, a major factor is certainly the need for *partitioning of energy*. Oxidation of glucose (5 mM) by oxygen (0.23 atmospheres) will produce large amounts of energy (2800 kJ/mol).

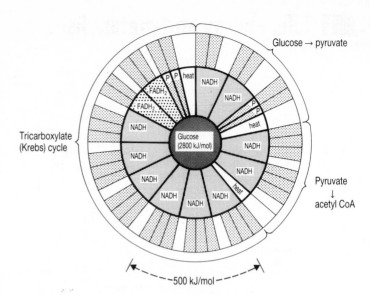

Fig. 1.2. Partitioning of energy from glucose (centre) into reduced cofactors (inner wheel) and ATP by oxidative phosphorylation (outer wheel). P = ATP, unshaded regions = energy dissipated as heat. *Note that extramitochondrial NADH (upper/right quadrant) yields only 2 mol ATP/mol.*

Biosynthetic reactions, in contrast, require much smaller amounts—formation of one peptide bond from two amino acids (1 mM) in water requires only 25 kJ/mol. Thus catabolism is split up into many steps, such that the energy released at each step is an amount convenient for the organism to handle. Convenient amounts seem to be 200 kJ/mol or less—NADH (10 yielded per glucose oxidized) traps about this amount of energy, reduced flavin (2 per glucose) traps about 140 kJ/mol, and ATP (4 per glucose) traps about 55 kJ/mol (Fig. 1.2). Note that energy from the oxidation of NADH and reduced flavin can be further partitioned into ATP.

Efficiency of energy conservation

If we compare the amount of energy released by oxidizing glucose to the amount trapped as NADH, reduced flavin and ATP, we find that a significant quantity of energy is 'lost' (Fig. 1.2). Catabolism is *less than 100 per cent efficient* at trapping energy.

Nonetheless, this 'lost' energy should not be thought of as 'wasted'. First it appears in the cell as generated *heat*, helping to maintain a body temperature above that of the surroundings (and thus to speed up biochemical reactions). Secondly, it ensures that the overall pathway is energetically downhill, and hence maintains a *flux* through the pathway in the right *direction*. (Otherwise, the pathway would simply run down to equilibrium, and stop.) Thirdly, because the precise energy change for a given reaction varies with conditions—especially with reactant concentration (Section 4)—it provides a cushion which ensures that even under the least favourable physiological conditions likely to be encountered (low [glucose], high [NADH]), *the arrangements for energy partitioning can still function with the same stoichiometry*. In figurative terms, even if the 'price' of NADH rises temporarily, energy income is still sufficient to 'buy' the same number of molecules.

Non-equilibrium reactions in metabolism

If we consider where, in metabolic pathways, this energy is 'lost', we find that dissipation occurs essentially at only a few, particular, reactions. These are the reactions *far displaced from equilibrium*. The remaining reactions—the majority of those in any metabolic pathway—operate close to equilibrium in the cell, and at these little energy dissipation occurs. (For the precise relationship between equilibrium constant and energy change, see Section 4.)

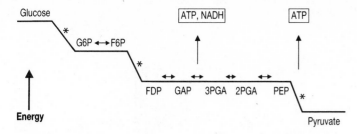

Fig. 1.3. Energy profile of glycolysis. The asterisks indicate non-equilibrium steps.

In the case of glycolysis (Fig. 1.3), for example, hexokinase, phosphofructokinase and pyruvate kinase catalyse non-equilibrium reactions. It is these reactions that determine the direction of the glycolytic flux; their reversal requires energy input, while the remaining, *near-equilibrium* reactions are, by definition, readily reversible.

One important feature of this arrangement is that **control** *can be exerted only at non-equilibrium steps.* Since, at equilibrium, the [product]/[reactant] ratio of an enzyme is fixed, slowing it down cannot affect the flux through a pathway since the other enzymes cannot 'sense' the change. In contrast, slowing down a non-equilibrium (undirectional) enzyme will decrease the concentration of its product, which will in turn slow down utilization of this product and thus subsequent steps in the pathway. Thus, to control metabolic pathways, we need non-equilibrium (energy-dissipating) steps, and this gives us a *fourth* reason as to why metabolic pathways operate at less than 100 per cent efficiency as regards energy trapping.

Note that classification of metabolic reactions into equilibrium and non-equilibrium types does not have implications as to *energy conservation*—whether ATP and/or NADH are generated. Energy conserving reactions may be either near-equilibrium or non-equilibrium *in vivo*. In glycolysis, oxidation of glyceraldehyde-3-phosphate to 3-phosphoglycerate is coupled to ATP and NADH production (Section 8), and occurs close to equilibrium. Conversion of phosphoenol pyruvate to pyruvate generates ATP, but also releases excess energy. The requirement for energy conservation is that the reaction *in the absence of the energy trap* (e.g. if glyceraldehyde-3-phosphate were oxidized directly by oxygen) would be highly energy yielding. Part or all of the energy yielded may then be trapped, depending on the available mechanism.

2 Energy, entropy and the universe

The first law of thermodynamics

The first law of thermodynamics states the equivalence of **heat**, **work** and **energy** in quantitative terms. If a mass, m, is raised to a height, h, the amount of work done is mgh joules; the increase in its *potential energy* is mgh joules; and the amount of heat released if it were dropped also should be mgh joules (although this may be written, in different units, as $mgh/4.2$ calories; Fig. 2.1). This law also emphasizes that work output is impossible without energy input; even if we could convert all potential energy into work, we could not exceed a work output of mgh joules.

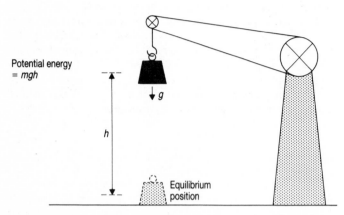

Fig. 2.1. Mechanical illustration of potential energy and equilibrium.

The second law of thermodynamics

The second law of thermodynamics determines the direction in which changes occur. Originally developed to deal with heat flows, it tells us in one formulation that heat tends to move from one body at a high temperature to another at a lower temperature—and *not* vice versa. Note that the first law does not deal with the problem of direction—it simply affirms that heat lost by one body must equal that gained by the other, independently of the direction in which the transfer occurs.

This principle can be restated in more general terms viz. that *the entropy of the universe tends to a maximum (i.e. any change in the universe will involve an increase in its entropy)*. At first sight, this seems to be a less helpful formulation—what is this '**entropy**' that dictates the direction of changes and, in any case, how could we possibly measure entropy changes in the whole universe?

Entropy and equilibrium

To get some idea of what entropy is, we can consider more directly how 'heat transfer' occurs at the molecular level. In an assembly of molecules, a fixed amount of energy is distributed between different sorts of molecular movements e.g. rotations, translation etc. at a given temperature. This energy, we know, is not restricted to only a few molecules ('hot molecules'); nor, indeed, does every molecule have the same energy. Instead, the energy is partitioned between the molecules in the statistically most probable manner, the **Boltzmann distribution** (Fig. 2.2b).

If we could, somehow, produce a situation where only a few, very 'hot' molecules carried all the energy (Fig. 2.2a), it would be unstable; the system would change towards, and finally reach, the Boltzmann distribution of energies. Then no further change would occur—the system is at **equilibrium**.

The second law of thermodynamics tells us that this change occurs because entropy increases. We can see why it occurs,

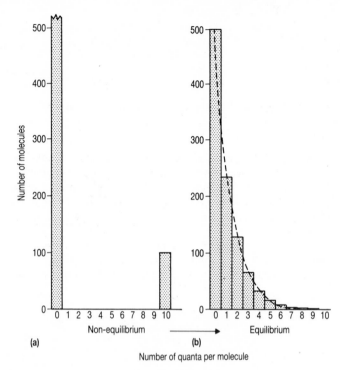

Fig. 2.2. (a) Very improbable distribution of 1000 quanta of energy between 1000 molecules.

(b) Most probable distribution of 1000 quanta of energy between 1000 molecules. The dotted curve is given by

$$N_i = N_0 \exp(-e_i kT)$$

that is, by the Boltzmann distribution.

mechanistically; there is a change towards the *most probable* energy distribution between the molecules. Thus entropy is a measure of *probability*, and the statement that the entropy of the universe tends to a maximum becomes more meaningful (and perhaps more obvious) if restated as the tendency of the universe to *attain the most probable distribution of its energy*. Importantly, we also see that **equilibrium is a condition of maximum entropy**.

Steady states

The picture painted above is rather a bleak one; ultimately heat (and mass) transfer will ensure that everywhere in the universe has the same temperature and composition. There will be no areas of specialization or 'order' in the universe—it will have attained its 'heat death'.

While these ideas are logical extensions of the above arguments, they seem to contradict our everyday experience. The Earth, for example, is neither the same composition, nor the same temperature, as deep space—and over fairly long periods, has not changed towards them. This is because the Earth is at a **steady state**.

Unlike an equilibrium, *a steady state must be maintained by continuous input of energy*. In the case of a planet, heat is continuously lost by radiation to the lower temperature of space. To maintain a steady state, this needs to be replaced by *energy input*—in fact, by the absorption of solar radiation (Fig. 2.3a). In accordance with the first law, the energy absorbed must exactly equal the energy lost; in accordance with the second law, heat is moving from the (high temperature) sun to the (medium temperature) planet, and on to (cold) space—and the entropy of the universe increases.

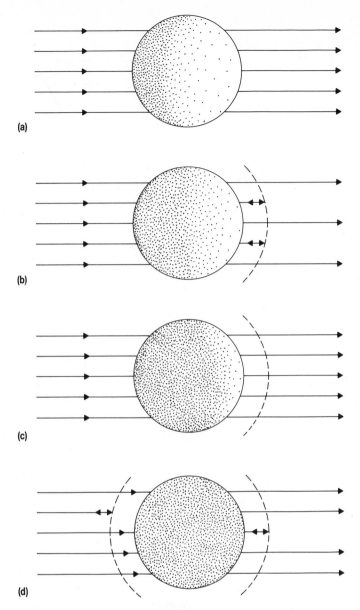

Fig. 2.3. Steady state as a balance between energy input and output.
  (a) Steady state: absorption = re-emission.
  (b) Steady state perturbed: re-emission prevented.
  (c) Steady state re-established at higher mean temperature.
  (d) Steady state: energy flow in planet is rapid relative to absorption/re-emission.
(*Increased hatching = increased temperature*)

Again, unlike an equilibrium, the nature of a steady state is directly affected by **kinetic** factors. Taking the Earth as an example, if the rate of heat loss were slowed down, e.g. by increasing CO_2 levels in the atmosphere, the mean temperature would rise (the 'greenhouse effect') until the new rate of heat loss equalled the energy absorption rate, and a new (different) steady state would be established (Fig. 2.3c).

Note that if no heat (or mass) loss from the planet was possible, it and the sun would come to the same temperature—an equilibrium would be achieved and no more energy transfer (or entropy loss) would be possible. More realistically, if the kinetic barriers to energy absorption and loss by the Earth were very high (energy fluxes were very low), but energy transfer around it was very rapid, the planet would reach a uniform temperature everywhere. This is roughly the case on Earth (Fig. 2.3d) as compared to, say, Mercury (Fig. 2.3a). An observer on the Earth thus sees locally, a *situation very close to equilibrium*. This situation, where a global steady state allows (using suitable kinetic barriers) local equilibria is particularly applicable to the situations we will consider in the remainder of this text.

Can life exist?

A living organism, too, represents a system out of equilibrium with its surroundings. Not only is your temperature higher than 'room temperature', your body contains compounds which are *unstable under bodily conditions*. Proteins, for example, are made by unfavourable dehydration reactions in an aqueous environment, and in time would hydrolyse in cell water. Equilibrium—the long term absence of change—is death.

Maintaining life, therefore, means maintaining a steady state. Energy input is required; in animals this comes from **chemical reactions**. For example, glucose is far from being in equilibrium with oxygen—and energy is released when it is oxidized. The extension of the above theoretical rationale from physical processes to such chemical ones will occupy the next few sections of this book.

As has been shown above, lack of overall equilibrium in a steady state does not preclude local equilibria; these depend on kinetic factors (relative rates). If you wear a coat to slow down bodily heat loss, your skin remains close to body temperature. Less obviously, glucose-6-phosphate and fructose-6-phosphate, intermediates in glucose oxidation, are *nearly at equilibrium within the cytoplasm* because their interconverting enzyme, phosphohexoisomerase, works much faster than some other enzymes in the oxidation pathway. The identification of non-equilibrium (potentially energy dissipating) reactions in the cell is another feature of the discipline of bioenergetics.

Thermodynamic principles, therefore, are quite consistent with the maintenance of living systems—although they do set certain constraints on the models we can use. The *appearance* of life, or its evolution from simple chemicals, is more problematic. Luckily for us, the process is at least feasible. Clearly, energy is needed to drive unfavourable reactions—for example, sunlight evaporating a rock pool could drive a dehydration. The other, less obvious, factor is again a kinetic one. This new compound needs to be stable enough to remain, even after the tide has come back in, to take part in another energy requiring process, and another, and so on *by chance* until it becomes able to *harvest energy and replicate itself*. The problem here is one of time—for how long does the compound need to be stable, what is the chance of the next combination, etc. In thermodynamic terms, evolution moves through a series of increasingly unlikely—but kinetically stable—steady states. Evolution of life is feasible but improbable; how improbable we can decide only when a statistically significant number of inhabited and uninhabited planets is available for study!

3 Energy, entropy and the living cell

Entropy and chemical reactions

Chemical processes proceed with breakage and re-formation of chemical bonds—and thus energy and entropy will alter during these changes. The second law of thermodynamics—designating the direction of change—tells us that:

1 $A + B$ will tend to change into C so long as there is a net increase in the entropy of the universe.

2 A mixture of A, B and C will reach equilibrium when the entropy of the universe is maximized, i.e. when converting $A + B$ to C OR C to $A + B$ would result in an entropy *decrease*, and

3 A steady state can be maintained away from equilibrium providing that EITHER $A + B$ are replaced and C is removed at the same, constant rate (*common in metabolic pathways within cells*) OR $A + B$ change infinitely slowly into C (*there is a kinetic barrier to the conversion*).

However, although we have a reasonable idea of the nature of entropy, the above statements are not predictively useful in themselves; it is difficult even to conceive of estimating entropy changes in the entire universe.

Potential energy and chemical reactions

We saw in Section 2 that a term, **potential energy**, governed the direction of a mechanical change. It would be useful to have a general 'potential energy' function applicable to all processes, even chemical reactions, of the same nature as mechanical potential energy. This function must relate, somehow, to the entropy of the universe, but should ideally obviate our need to measure the latter. This section shows that such a function does indeed exist—it is denoted G—and will examine its properties.

Consider a ball rolling down a slope ('potential energy surface') as in Fig. 3.1a. Wherever it is placed, it will come to rest at position A, the position of least potential energy ('equilibrium position'). An analogous diagram for a chemical process is given in Fig. 3.1b. The non-spherical nature of the 'ball' in this case indicates a 'kinetic barrier' which limits the rate of attaining equilibrium.

As the ball rolls (Fig. 3.1a), its potential energy falls by $mg\Delta h$ (joules), where Δh is the *vertical* distance traversed. While the ball is moving, this energy is distributed between kinetic energy and heat (due to friction); at rest, at A, all the potential energy has been converted to heat ($= mg\Delta h$ joules). The ball itself warms up slightly, but ultimately regains its original temperature since the *heat is transferred to its surroundings*. It is in this final process that we see the *connection between loss of potential energy and gain in entropy of the universe*—the potential energy is converted to heat which, flowing from the warm ball to the cooler surroundings, increases the entropy of the universe.

A chemical reaction is more complex—changes in atomic associations, and composition of the mixture, mean that entropy changes occur both within the mixture under study (ΔS_{system}) and in its surroundings ($\Delta S_{\text{surround}}$). However, it is still true that an increase in entropy of the surroundings is the result of *heat flow out of the system*.

Mathematically.

$$\Delta S_{\text{surround}} = f(\Delta H_{\text{system}})$$

(where ΔH represents the heat removed from the system to keep the temperature constant).

In fact,

$$\Delta S_{\text{surround}} = -\Delta H_{\text{system}}/T$$

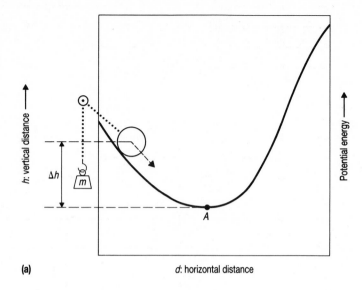

(a) *d*: horizontal distance

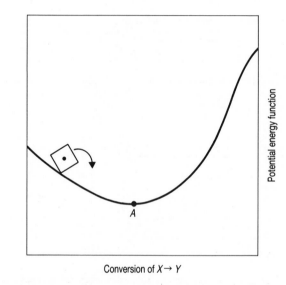

(b) Conversion of $X \rightarrow Y$

Fig. 3.1. **(a)** Potential energy surface for rolling ball.
(b) Potential energy surface for chemical reaction.

where T is absolute temperature.* Thus

$$\Delta S_{\text{universe}} = \Delta S_{\text{system}} + \Delta S_{\text{surround}}$$

$$= \Delta S_{\text{system}} - \Delta H_{\text{system}}/T$$

We define our 'potential energy function', G, such that, for ANY change

$$\Delta G = \Delta H_{\text{system}} - T\Delta S_{\text{system}} = -T\Delta S_{\text{universe}}$$

To summarize:

1 We have replaced $\Delta S_{\text{universe}}$ by an equivalent function requiring measurements to be made only on a system under study rather than on the entire universe. (Suffixes relating to *system* will be omitted in subsequent sections.)

* The negative sign indicates that, for heat *lost* from the system, $\Delta S_{\text{surround}}$ *increases* (more energy quanta are distributed about the surroundings). The $1/T$ factor is necessary because a given amount of heat will increase entropy more (create a relatively larger change in molecular energy distribution) if very little energy (very few quanta) was originally present. See the definition of entropy, above.

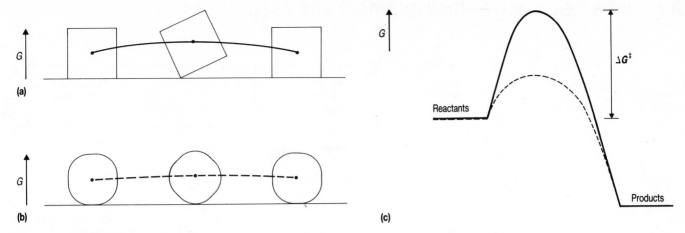

Fig. 3.2. (a) Path of centre of gravity of square 'ball'.
(b) Path of centre of gravity of rounded 'ball'.
(c) Conventional representation of (a) and (b) in terms of 'activation energy' of reaction.

2 *At equilibrium*, as $S_{universe}$ must be maximal, **G must be at a minimum** like the mechanical potential energy in the analogy.

Maximum available work

If we consider again our mechanical analogy (Fig. 3.1a), in the absence of any external constraints, the energy available was converted into heat—the heat being equivalent to $mg\Delta h$, where Δh is the (vertical) difference between the initial position of the ball and its final position (*A*). It would be feasible to use the potential energy of the ball to do mechanical work, e.g. by connecting it, via a pulley, to raise a weight (shown dotted). In this case, from the first law of thermodynamics:

Total work done = potential energy change −
energy lost as heat (by friction)

We considered above the case when no work was done—all the energy appeared as heat. Conversely, in a frictionless system,

Total work done = potential energy change

since, in this case, no energy appears as heat.

The important conclusions from this argument are that:

1 *The potential energy change from initial position to equilibrium = maximum work available from a system.*

2 *If the system is initially at equilibrium*, not only will it not tend to change, but *no work can be obtained from it.*

3 Normally the work obtained from a system is less than the maximum work available, as some energy is lost to warm up the surroundings.

All these conclusions hold in the wider case of our potential energy (or **free energy**) function, G, and they thus apply to chemical as well as physical processes. In chemical processes, the equivalent to a pulley (converting the energy released to the required work) is more generally termed a **coupling device**—typically a protein (or complex of proteins) in biochemistry.

Rate of reactions

The parameter, G, allows us to predict the *direction* of a chemical reaction. If ΔG is negative, as $A + B \rightarrow C$, then the reaction will proceed in this direction; if it is positive, there will be a tendency for C to be converted to $A + B$. The only factor affecting potential energy in our mechanical analogy was the height, 'Δh'. Many more factors affect G for a chemical change, as we shall see in the next section.

However, as noted above, kinetic restraints will affect the rate of any potential energy change, and this is represented in Fig. 3.1b by the non-spherical shape of the 'ball'. Thus, ΔG values do not determine the rates of change—changes may take place infinitely slowly. Change is restricted by **activation energy**, represented very loosely in Fig. 3.2a by the path of the centre of gravity of the rolling 'ball' ($\Delta G\ddagger$ in Fig. 3.2c).

While the nature and treatment of activation energies are interesting, they are not relevant in this context—particularly since, if a reaction is required in biology and has a negative ΔG, an enzyme will normally evolve to CATALYSE (decrease the activation energy of) the reaction. This is represented in Fig. 3.2b by 'rounding off' the corners of the brick.

4 Gibbs free energy—measurement and applications

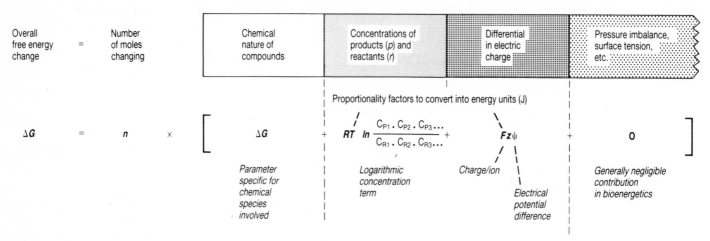

Fig. 4.1. Components of chemical free energy changes.

Parameters affecting G

Our 'potential energy function', G (known as the Gibbs free energy) is, in principle, very useful; it allows us to predict the likely *direction* of chemical (and physical) changes, and to estimate the *amount of work* we can maximally extract from a system. Strictly, this is true only for systems maintained at constant temperature by transfer of heat in and out (*isothermal, closed systems* in thermodynamic jargon) but to a first approximation, these conditions approximate to many biochemical processes.

So far, however, we cannot apply the parameter G, since it is not clear from above how to measure it. To solve this problem, we will consider those factors which must affect changes in 'potential energy' (or, direction of change) in a generalized chemical process.

Type of compounds involved in the change

The direction of reactions, by experience, depends on the type of compounds involved. For example, reaction (a) will tend to move to the right (i.e. ΔG must be negative in the direction of the arrow) while the formally similar reaction (b) will tend to move to the left.

(a) $H_2 + \frac{1}{2}O_2 \rightarrow H_2O$ ΔG *(left to right)* negative

(b) $N_2 + \frac{1}{2}O_2 \rightarrow N_2O$ ΔG *(left to right)* positive

Thus, an expression giving ΔG must contain a term which relates to the type of compounds involved in the change. The value of this term is not obvious simply from inspection of the reaction equation, and it is denoted ΔG^0.

Concentration of compounds present

The direction of a given reaction, by experience, depends on the concentration of reactants. In the reaction

$(glycogen)_n + P_i \rightleftharpoons (glycogen)_{n-1} + glucose\text{-}1\text{-}phosphate$

if the concentration of P_i is very high, and of G-1-P is low (and a suitable enzyme catalyst is present), glycogen is phosphorylysed (*left → right*). Conversely, if the concentration of G-1-P is high, and that of P_i low, the reaction is reversed. Thus our parameter of reaction direction, ΔG, must contain terms relating to **reactant concentration** (c_R) and **product concentration** (c_P).

Extent of reaction

The energy change, in any process, will depend on the amount of material participating. To raise a heavy ball a given distance requires more work than to raise a lighter ball the same distance. Similarly, ΔG for hydrolysis of 1 mol ATP must be $1000 \times \Delta G$ for hydrolysis of 1 mmol ATP, under the same conditions.

Electrical potential

Ions (charged atoms or molecules) will move towards an electrode of opposite charge. In other words, the direction of ion movement is influenced by the *charge on the ion*, z, and the *electrical potential difference*, ψ, between its position and somewhere else in the solution. For charged molecules in an electric field, therefore, both z and ψ must contribute to the potential energy term, G.

Summary

Summarizing all these four effects, and allowing for the (here unproven) assumption that the dependence of ΔG on concentration is *logarithmic* while those on electrical potential and reaction extent are *linear*, we have the equation given above (Fig. 4.1), here rewritten (using Π to denote a multiplication product)

$$\Delta G = n(\Delta G^0 + RT \ln \Pi c_P / \Pi c_R + zF\psi) \qquad \text{(eqn 4.1)}$$

If we were to consider *all possible* chemical and physical processes, other factors would contribute to ΔG; ΔG for the falling ball would include the mechanical potential energy, $mg\Delta h$ (see Section 2). However, if we consider just these four contributions, we are equipped to deal with most situations in bioenergetics.

Properties of ΔG

Considering first a chemical reaction in the absence of electric field effects (z or $\psi = 0$), equation 4.1 has important consequences.

1 n, the number of moles reacted, will influence the magnitude of the energy change (amount of work possible) but not its sign (rather like m, the mass of the ball in our mechanical analogy). It is often convenient to consider $\Delta \overline{G}$, the free energy change per mol, in predicting the *direction* of a chemical process.

2 If the concentrations of reactants (c_R) or products (c_P) change as a reaction proceeds, an overall ΔG can be calculated only as an integral function over the total change. However, if the concentrations are unchanged in the reaction, this simplifies to

$$\Delta \overline{G} = \Delta G^0 + RT \ln \Pi c_P / \Pi c_R \qquad \text{(eqn 4.2)}$$

Luckily, homeostasis (replacement of reactants, and removal of products) is usual in biological systems, allowing us to use the simpler expression in most cases.

3 The free energy released in a given reaction depends not only on the types of compound reacting but also on their concentration. In the reaction.

$$ATP + H_2O \rightarrow ADP + P_i$$

ΔG at 10 mM ATP, 0.1 mM ADP, 10 mM P_i and 55 M water (i.e. the conditions prevailing in the cell) is about −55 kJ/mol (designated ΔG_P, the cellular **phosphorylation potential**)—but at 1 M ATP, 1 M ADP, 1 M P_i and 55 M water, ΔG would be only −30 kJ/mol (see Fig. 4.2). This seems intuitively odd; one might ask how one ATP molecule 'knows' the amount of energy to release if this depends on the concentration of surrounding molecules. This paradox vanishes if we remember that ΔG is a function corresponding to potential energy. A ball does not 'know' how much potential energy it holds and how much to release in coming to equilibrium; this too depends on its surroundings—the position of a plane external to the ball itself.

4 The sign of ΔG depends on the direction of reaction. If ΔG is negative for the reaction $A \rightarrow B$ (i.e. A will tend to be converted to B), it must be equal but positive (under the same conditions) for the conversion of $B \rightarrow A$. Thus, *in giving ΔG for a reaction, its direction must be specified.*

5 The equilibrium between products and reactants is represented by the trough in our energy diagram (Fig. 4.2). By definition, to *move away from equilibrium*, in either direction, work must be done on the system; ΔG *is positive*. For conversions which do not involve moving out of the well—reactions under equilibrium conditions—ΔG must equal zero.

Thus our criteria of equilibria are:

$S_{universe}$ = maximum

or G = minimum

or ΔG = 0

No work can be obtained from a reaction taking place at, or very close to, equilibrium conditions.

In the cell, many reactions take place close to equilibrium—for

example the conversion between glyceraldehyde-3-phosphate and dihydroxyacetone phosphate in glycolysis (see Section 2).

Equilibrium constants

Equation 4.2 provides us with a method for measuring ΔG^0, the parameter that was stated above to depend on *reaction chemistry*. We have

$$\Delta \overline{G} = \Delta G^0 + RT \ln \Pi c_P / \Pi c_R$$

At equilibrium

$$\Delta G = 0 \quad \text{(above)}$$

and the concentrations reach their final (i.e. equilibrium) values, $c_{P1}{}^{eq}, c_{P2}{}^{eq}, \ldots$ etc.

Then

$$0 = \Delta G^0 + RT \ln \Pi c_P{}^{eq} / \Pi c_R{}^{eq}$$

or

$$\Delta G^0 = -RT \ln \Pi c_P{}^{eq} / \Pi c_R{}^{eq} \qquad \text{(eqn 4.3)}$$

Since ΔG^0 is constant for the given reaction, and R and T are constants under given reaction conditions, then

$$\Pi c_P{}^{eq} / \Pi c_R{}^{eq} = K_{eq},$$

which is the familiar *equilibrium constant* for a reaction. Because we can measure equilibrium concentrations, we can calculate the value of K_{eq}, and $-\Delta G^0$ ($= RT \ln K_{eq}$).

Standard conditions

ΔG^0 can be considered as the free energy change when (1 mol) reactant is turned into product under a given set of standard conditions. This can be seen by combining equations 4.2 and 4.3.

$$\Delta G = n(-RT \ln \Pi c_P{}^{eq} / \Pi c_R{}^{eq} + RT \ln \Pi c_P / \Pi c_R) \qquad \text{(eqn 4.4)}$$

As we have seen

$$\Delta G = 0 \quad \text{when } \Pi c_P / \Pi c_R = \Pi c_P{}^{eq} / \Pi c_R{}^{eq},$$

i.e. at equilibrium.

In addition, equation 4.4 indicates that

$$\Delta \overline{G} = \Delta G^0 \quad \text{when } \Pi c_P / \Pi c_R = 1 \quad (\ln \Pi c_P / \Pi c_R = 0)$$

c_P and c_R are measured relative to standard values (standard conditions). In biochemistry, the standard conditions commonly taken are:

[all reagents] = 1 mol/l (1 M)* except

[water] = 55 M (i.e. pure water)

[H$^+$] = 10^{-7} M

[gases] = 1 atmosphere pressure

ΔG^0 for these standard conditions (neutral, aqueous solution) is written $\Delta G^{0'}$. Figure 4.2 shows that ΔG must vary as conditions depart from this standard, and also that the value of ΔG^0 must vary if different standard conditions are chosen.

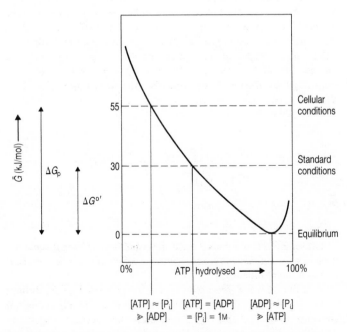

Fig. 4.2. Free energy schematic for ATP hydrolysis at pH 7.0 under standard and non-standard conditions (cf. Fig. 3.1).

* The convention of taking 1 M as a standard concentration is common in chemistry. It may not be the best standard for the biochemist—cellular concentrations of organic compounds are typically around 1 mM. This means that directions of reactions *in vivo* cannot be simply determined by inspecting the value of $\Delta G^{0'}$.

◼ 5 Chemical and electrochemical potential

Chemical potential

From Section 4, we know that ΔG for the (physical or chemical) process $A \to B$ (in the absence of any electrical charge imbalance, $\psi = 0$) is given by

$$\Delta G_{A \to B} = n(\Delta G_{A \to B}^{0'} + RT \ln c_B/c_A)$$

If we arbitrarily split $\Delta G_{A \to B}$ into contributions from the two species A and B, and denote these by μ_A and μ_B respectively, we have

$$\Delta G_{A \to B} = n(\mu_B - \mu_A) \text{ as } A \text{ disappears and } B \text{ appears} \qquad \text{(eqn 5.1)}$$

Splitting μ_A and μ_B further

$$\Delta G_{A \to B} = n[\mu_B^0 + RT \ln c_B) - (\mu_A^0 + RT \ln c_A)] \qquad \text{(eqn 5.2)}$$

μ_A is known as the **chemical potential of** A. This is convenient, since if $\mu_A > \mu_B$, $\Delta G_{A \to B}$ will be negative, or, in words, *if the chemical potential of A is greater than that of B, A will tend to change into B.*

Chemical potential is thus analogous, for a chemical compound, to the mechanical potential (energy) of Section 2; a system will react so that the compound at high chemical potential will change into another at lower chemical potential. At equilibrium, the chemical potential of both species will be equal; $\mu_A = \mu_B$, or $\Delta G_{A \to B} = 0$.

This is seen, in its simplest possible form, in Fig. 5.1. Compound X is introduced in the region of A, at a concentration $[X]_A$. There is no chemical change, so $\mu_A^0 = \mu_B^0 (= \mu_X^0)$. However, since its chemical potential is concentration dependent, X will be at higher potential at A than at B and, if a kinetic pathway is available (diffusion here), X will move from A to B until its concentration (and thus its chemical potential) is equal everywhere.

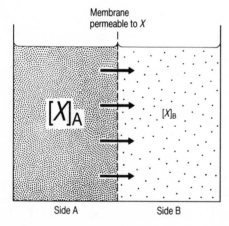

Fig. 5.1. Distribution of an uncharged solute. No chemical or charge change; equilibrium only when $[X]_A = [X]_B$.

Systems with more than one reactant or product

The above represents a simple, physical change in which no chemical reaction occurs. Its extension to a chemical change, which has several reactants $R_1, R_2, \ldots R_n$, and several products, $P_1, P_2 \ldots P_n$, is straightforward. In this case, from 4.1.

$$\Delta G_{R \to P} = n(\Delta G_{R \to P}^{0'} + RT \ln c_{P1}, c_{P2} \ldots c_{Pn}/c_{R1}, c_{R2} \ldots c_{Rn})$$

But, from equation 5.2, $\mu_A = \mu_A^0 + RT \ln c_A$

Thus equilibrium occurs when $\Delta G_{R \to P} = 0$

$$\mu_{R1} + \mu_{R2} + \ldots \mu_{Rn} = \mu_{P1} + \mu_{P2} + \ldots \mu_{Pn} \qquad \text{(eqn 5.3)}$$

$$(\text{or } \Sigma\mu_{Ri} = \Sigma\mu_{Pi})$$

Electrochemical potential

In a process where charge movement is involved, we can no longer ignore the electrical contribution to potential energy, and must use the complete expression for ΔG (equation 4.1). Again considering the process $A \to B$,

$$\Delta G_{A \to B} = n(\Delta G_{A \to B}^{0'} + RT \ln c_B/c_A + zF(\psi_B - \psi_A)) \qquad \text{(eqn 5.4)}$$

(where $(\psi_B - \psi_A)$ represents the electrical potential at B relative to A, simply written as ψ above.)

We can again write

$$\Delta G_{A \to B} = n(\mu_B - \mu_A)$$

(where equilibrium is given by $\Delta G_{A \to B} = 0$ or $\mu_A = \mu_B$) but now an additional electrical component enters our definition of μ

$$\mu_A = \mu_A^0 + RT \ln c_A + zF\psi_A \qquad \text{(eqn 5.5)}$$

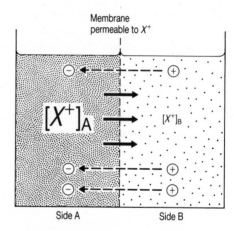

Membrane permeable to X^+

$[X^+]_A$ $[X^+]_B$

Side A   Side B

Fig. 5.2. Distribution of a permeable cation (impermeant anion). Equilibrium when tendency to move down concentration gradient (bold arrows) is balanced by charge gradient (dotted arrows).

We can now consider a perhaps more interesting equilibrium (Fig. 5.2). If X^+ is present at unequal concentrations *across a membrane permeable only to X^+* (not to its counterion), X^+ will tend to diffuse down its concentration gradient ($c_A > c_B$). However, in doing so, it will create excess positive charge on side B, leaving a negative charge on side A—and a voltage difference ($\psi_B - \psi_A$) will be created, opposing the diffusion. At equilibrium, the *concentration and voltage differences must just balance.*

At equilibrium

$$\mu_A = \mu_B$$

Again there is no chemical change, so $\mu_A^0 = \mu_B^0 (= \mu_X^0)$

$$RT \ln c_B/c_A = zF(\psi_B - \psi_A)$$

or

$$(\psi_B - \psi_A) = (RT/z/F) \ln c_B/c_A \qquad \text{(eqn 5.6)}$$

Equation 5.6 is sometimes termed the **(equilibrium) Nernst equation**; it relates the *equilibrium transmembrane potential to the ion gradient* across a membrane.

Note that very few ions need to move to establish this equilibrium; c_A and c_B, the equilibrium concentrations are essentially the initial concentrations at A and B. To emphasize the electrical term in the definition of μ_A in equation 5.5, it is often written as $\tilde{\mu}_A$, and called the **electrochemical potential**. This is a convenient term in bioenergetics since, as stated above, the changes we will consider involve changes in ΔG due almost entirely to these three factors—change in chemical nature (μ^0), concentration change ($RT \ln c$) and movement of charge in a field ($zF\psi$).

Other factors

In general the chemical potential (as arbiter of equilibrium) must contain *all* energy terms that might influence the direction of a process (see equation 4.1). A contrasting example is shown in Fig. 5.3, where equilibrium is maintained at unequal concentrations of water across a semipermeable membrane by applying a pressure ('osmotic pressure') to the low concentration side, *B*. In this case, the normally negligible pressure component of ΔG (see equation 4.1) becomes significant.

Thus, *for water*,

$$\mu_A = \mu_A{}^0 + RT \ln c_A$$

$$\mu_B = \mu_B{}^0 + RT \ln c_B + V\Pi$$

(where Π is the applied pressure, and V, again, a proportionality constant).

Equilibrium occurs when $\mu_A = \mu_B$, or

$$\Pi = (RT/V) \ln c_A/c_B \tag{eqn 5.7}$$

from which osmotic pressure can be calculated.

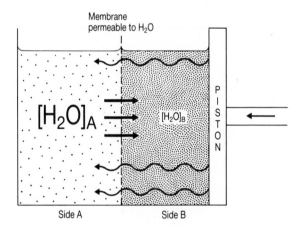

Fig. 5.3. Osmotic equilibrium. High solute concentration (low [H₂O]) on side B. Equilibrium when tendency to move down concentration gradient balanced by pressure.

Non-equilibrium situations

When $\mu_A \neq \mu_B$, a system is not at equilibrium and will change with a decrease in free energy (ΔG negative). One such situation is shown in Fig. 5.1. Another is shown in Fig. 5.4. Comparing this with Fig. 5.2, we see that in this case a balance between concentration and electrical forces cannot be achieved as both parameters promote flow in the same direction (here, left to right). Rather than cancelling out, the *concentration and electrical terms in the expression*

$$\tilde{\mu} = \mu^0 + RT \ln c + zF\psi$$

both add to the magnitude of ΔG.

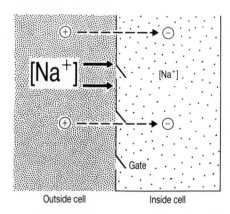

Fig. 5.4. Non-equilibrium distribution of differentially permeant cation. Na^+ moves down concentration and electrical gradient when gates open.

In this case

(a) ΔG for ion movement (left → right) is negative

i.e. ion movement can generate work (see Section 7), and

(b) ion movement will occur, unless restricted by a kinetic barrier.

This situation occurs in nerve cells, for example, which maintain a high $[Na^+]$, and a positive potential, outside. Normally, Na^+ movement is very slow, because the cell membrane acts as a kinetic barrier; however, when a nerve impulse passes, Na^+ channels ('gates') open, and Na^+ ions rush, downhill, into the cell to cause depolarization. A similar situation is observed, with respect to H^+ ions, in mitochondria and chloroplasts (Section 7), where ion movement is used to generate ATP.

6 Oxidation/reduction (redox) potentials

Redox potential

In **oxidation/reduction reactions**, chemical changes occur as a result of *changes in electron distribution*, which may, but need not, involve changes in atom distribution. As an example, we can consider the following reaction:

$$Fe^{II} + Cu^{II} \rightarrow Fe^{III} + Cu^{I}$$

The following conditions hold.

$$\Delta \overline{G} = \Delta G^{0'} + RT \ln (Fe^{III})(Cu^{I})/(Fe^{II})(Cu^{II})$$

(from eqn 4.2)

and

$$\Delta \overline{G} = (\mu_{Fe^{III}} + \mu_{Cu^{I}}) - (\mu_{Fe^{II}} + \mu_{Cu^{II}})$$ (from eqn 5,1).

If we (arbitrarily) regroup the chemical potential terms, we can rewrite this latter equation as

$$\Delta \overline{G} = (\mu_{Cu^{I}} - \mu_{Cu^{II}}) - (\mu_{Fe^{II}} - \mu_{Fe^{III}})$$

where, for example, the first pair of terms ($\mu_{Cu^{I}} - \mu_{Cu^{II}}$) *represents the chemical potential change during a hypothetical half reaction in which Cu^{II} is reduced.*

Directly derived from this hypothetical chemical potential change is the **redox potential**, E_{Cu}. Since electron transfers can be measured with electrodes (see Fig. 6.1), E_{Cu} may be measured electrically and is given in units of *electrical potential*, volts. If we represent ($\mu_{Cu^{I}} - \mu_{Cu^{II}}$) as equal to $F.E_{Cu}$, we can write

$$\Delta \overline{G} = (F.E_{Fe} - F.E_{Cu}) = -F.\Delta E$$ (eqn 6.1)

where F is the Faraday, the proportionality factor converting *volts* into energy units, *joules* (see Section 4).

Properties of the redox potential, E

1 E is conventionally taken to relate to the *reduction* direction of the half reaction; it can be *fully* specified by the symbol $E(ox/red)$ (e.g. $E(Fe^{III}/Fe^{II})$) although this is generally abbreviated E_x where there is no ambiguity (see below). The pair of reactants ox/red, differing only in electron number, is denoted a **redox couple**. For the above reaction as written, copper is reduced and iron is oxidized, and so

$$\Delta \overline{G} = -F(E_{Cu} - E_{Fe}), \text{ or}$$

$$\Delta E = E_{Cu} - E_{Fe}$$

(Note that, as stated before, ΔG has no conventional direction and thus requires the direction of reaction to be specified.)

2 E as measured relates to transfer of 1 mole of electrons; thus for a reaction in which z electrons are transferred per mol of reactant

$$\Delta \overline{G} = -z.F.\Delta E$$ (eqn 6.2)

(cf. ΔG, which refers to no conventional amount of change.)

An example of a reaction in which 2 electrons are transferred per mol reactant is

succinate + flavin → fumarate + flavin . H_2 (see Section 11).

In this case

$$\Delta \overline{G} = -2F.\Delta E$$

3 If the reaction tends towards reduction ($\mu_{red} - \mu_{ox}$ is *negative*), the measuring electrode will lose electrons to the reaction and become *positive*. Hence the negative sign relating ΔG to ΔE (equation 6.1).

4 We cannot study half a reaction. Ultimately all redox potentials are related by convention, to a standard hydrogen electrode

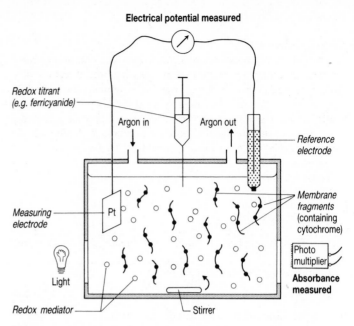

Fig. 6.1. Apparatus for measuring redox potentials of membrane cytochromes (schematic).

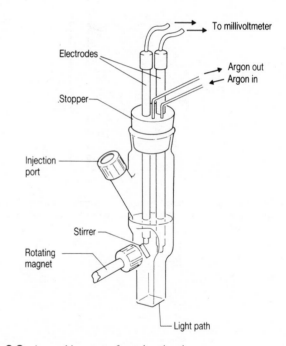

Fig. 6.2. Anaerobic cuvette for redox titration.

H_2 (1 atmos)/H^+ (1 mol/l) defined as possessing $E = 0$. (To indicate this, E may be written E_h.)

5 E is a measure of 'oxidizing power' or 'tendency to be reduced'; if E is positive, then the couple tends to take up electrons, i.e. it is a good oxidizing agent. Above, $E_{Fe} < E_{Cu}$, and so Cu^{II} is a better oxidizing agent than Fe^{III}; electrons pass from Fe^{II} to Cu^{II} (e.g. from cytochrome c to its oxidase).

(As an aid to memory, negative things turn electrons away; thus couples with a negative redox potential are good reducing agents.)

6 E is related to a parameter $E^{0'}$ (and is dependent on the chemical nature of the species involved) in much the same way as ΔG is related

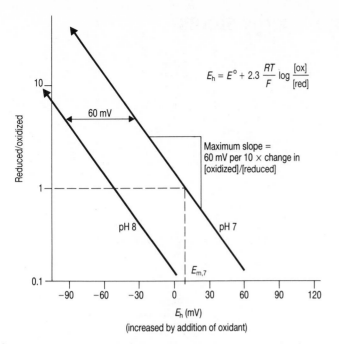

$$E_h = E^0 + 2.3 \frac{RT}{F} \log \frac{[ox]}{[red]}$$

Maximum slope =
60 mV per 10 × change in
[oxidized]/[reduced]

60 mV

pH 8

pH 7

$E_{m,7}$

E_h (mV)

(increased by addition of oxidant)

Fig. 6.3. Change in redox potential (E_h) with per cent of oxidation and pH. Reaction $AH \rightarrow A + e + H^+. E^{0'} = 10\,mV$

to $\Delta G^{0'}$. Allowing for the difference in units, for a transfer of z electrons (compare equation 4.2)

$$E = E^{0'} + (RT/zF) \ln [\text{oxidized}]/[\text{reduced}] \qquad (\text{eqn } 6.3)$$

If [oxidized] = [reduced], then $E = E^{0'}$. Since this value is reached when sufficient reductant has been added to reduce half of a given oxidant (i.e. at the **midpoint** of a titration), $E^{0'}$ is sometimes written E_m (see below).

7 The prime in $E^{0'}$, as in $\Delta G^{0'}$, represents a standard condition of neutral pH, pH 7.0. ($E^{0'}$ is sometimes written $E_{m,7}$ to emphasize this.) It is very common for oxidation reactions to liberate protons (see Section 10—in which case E will be dependent on pH as we move away from standard conditions. For example:

$$FADH_2 \rightarrow FAD + 2e + 2H^+$$

$$E = E^{0'} + (RT/2F) \ln [FAD][H^+]^2/FADH_2$$

Since RT/F has a fixed value, *increasing the pH by 1 unit* (decreasing [H^+] tenfold) *will decrease E by a fixed value*, – 60 mV at 37°C (in reactions which yield 1H^+/e removed), see Fig. 6.3.

Note that the standard hydrogen electrode has $E^0 = 0\,V$ at pH 0 (1 M H^+). Thus $E^{0'}$ for the hydrogen electrode will be $(-60) \times 7\,mV = -420\,mV$. Thus H_2 is a stronger reducing agent if [H^+] concentration is low compared to 1 M.

Measurement of redox potential of a cytochrome

Standard redox potentials of ion couples in free solution can be readily determined in *redox titrations*, in which the potential of a solution (E_h) is measured by an electrode while the extent of reduction of the ion ([oxidized]/[reduced]) is varied by adding a reductant or oxidant.

However, redox centres in proteins (i) are not readily accessible to electrodes placed in the solution; and (ii) usually only accept electrons from very specific (biological) donors. Thus *redox mediators* are used to carry electrons from the electrode and/or titrant to buried metal cofactors (e.g. haem). Note that equilibration may take some time—as even in these conditions, electrons can enter or leave the protein only slowly.

The experimental set-up is as in Figs. 6.1 and 6.2. A cuvette contains the cytochrome (soluble or in membrane fragments) at pH 7.0 in an oxygen-free environment. (Oxygen is a strong oxidant which would otherwise disrupt the titration.) The redox potential of the cytochrome is measured by the platinum electrode relative to the standard electrode, commonly Ag/AgCl/Cl$^-$ (which can be related to the hydrogen electrode). The redox mediator (phenazine methosulphate), in small amount (5 μM), can transfer electrons between the electrode, the titrant, and the protein, and is included.

The cytochrome is first fully reduced with the strong reducing agent, sodium dithionite ($Na_2S_2O_4$). Oxidation is induced by adding increasing amounts of ferricyanide through the injection port, and the oxidation state of the cytochrome is measured spectrophotometrically (see Section 10). An electrical potential is measured between the electrodes.

In general $E = E^{0'} + (RT/zF) \ln[\text{oxidized}]/[\text{reduced}]$.

When [oxidized] = [reduced] – the midpoint

then $E = E^{0'}$ ($E_{m,7}$)

Again, since RT/F is fixed, the slope of the graph

E *versus* log [reduced]/[oxidized] (Fig. 6.3)

will allow the determination of the number of electrons transferred, z. When Fe^{II} is oxidized to Fe^{III} (Fig. 6.3), $z = 1$ and thus the slope = 60 mV/tenfold change in reduced/oxidized (60 mV/decade). Figure 6.3 also shows the dependence of the titration on pH if 1 H^+/e is liberated during oxidation.

19

7 ATP and ion gradients: 'intermediate' energy stores

Intermediate stores of energy

Energy input, in living organisms, is from light or oxidation reactions—typically yielding upwards of 200 kJ/mol of substrate (or photons) used. Energy utilizing reactions (biosyntheses etc.) require 10–50 kJ/mol. To link these two processes together, organisms possess intermediate energy stores, or 'energy currency' which allows energy from the input reactions to be *partitioned into usable quantities*. These intermediates are phosphoric acid anhydrides, especially **adenosine triphosphate** (ATP) (Figs 7.1 & 7.2), and **transmembrane ion gradients** (Fig. 7.4).

ATP is the more generally useful, since it can diffuse around the cell and participate directly in chemical reactions in any cell compartment. Transmembrane ion gradients can be used with much less flexibility; they are limited to driving processes associated with the membrane involved—but they are conveniently employed in driving several membrane transport systems.

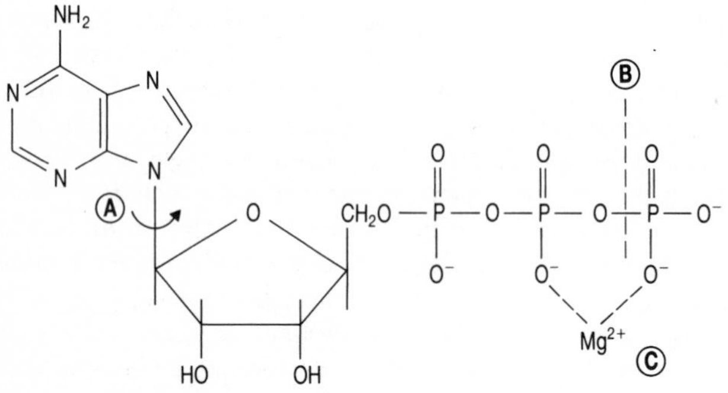

Fig. 7.1. Structure of ATP.
Ⓐ Site of rotation between anti (shown) and syn conformers.
Ⓑ Site of cleavage to give ADP.
Ⓒ Site of complexation with Mg^{2+}. Under cellular conditions, most ATP is present as the $[ATP\ Mg]^{2-}$ complex.

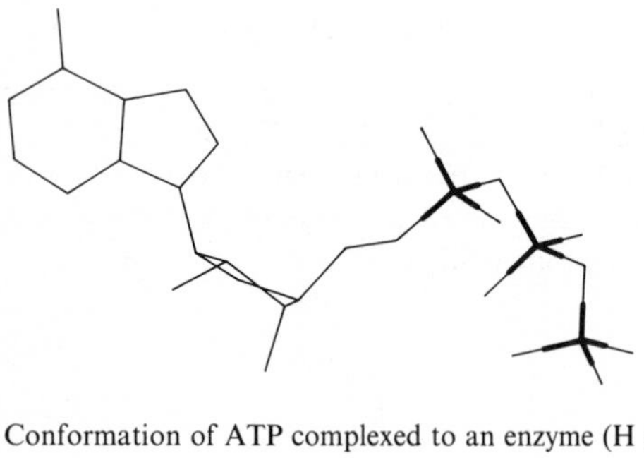

Fig. 7.2. Conformation of ATP complexed to an enzyme (H atoms not shown).

ATP—a diffusible energy store

From Sections 3–6, it is clear that *any compound maintained away from equilibrium may serve as an energy source*. The question then arises, why is the particular compound, ATP, employed *in vivo* almost exclusively? The essential features of ATP are:
1 its high negative $\Delta G^{0'}$ for hydrolysis (–30 kJ/mol), and
2 its kinetic stability in water.

High negative $\Delta G^{0'}$

Chemically, $\Delta G^{0'}$ for ATP hydrolysis is very negative because ATP is a (phosphoric) **acid anhydride** in water; like other acid anhydrides (e.g. acetic anhydride) its hydrolysis products are stabilized by

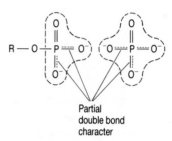

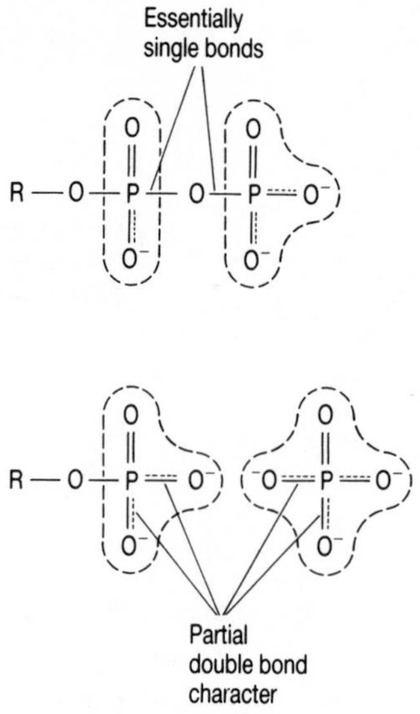

Fig. 7.3. Resonance systems for ATP and its hydrolysis products.

hydration and electron delocalization (Fig. 7.3). This value of $\Delta G^{0'}$ means that *ATP can be maintained far away from equilibrium at reasonable concentrations (0.1–10 mM) of reactants and products* (see below).

In contrast, glucose-6-phosphate (G6P) has a much less negative $\Delta G^{0'}$ for hydrolysis (–11 kJ/mol). Thus, while ΔG can reach a value of –55 kJ/mol, as for ATP, this could occur at a cellular $[P_1]$ of 10 mM, only if
(a) [glucose] $= 8 \times 10^{-8}$ M   if [G6P] $= 5$ mM, or
(b) [G6P] $= 6.1$ M (1200 g/l)   if [glucose] $= 0.1$ mM
In other words, to maintain one substrate or product at a reasonable level, the other must be either (i) so low as to severely limit reaction rates, or (ii) so high as to totally displace cell water!

Note that, under cellular conditions, ATP shows a more negative ΔG for hydrolysis (–50 to –55 kJ/mol) than under standard conditions (Section 4).

Kinetic stability

Once formed, in the absence of a catalyst, *ATP is stable in water for several days*. Clearly, if ATP were readily hydrolysed (kinetically unstable) like its arsenate analogue (ADP-As), its energy would be rapidly dissipated and it would be useless as an energy store. Thus, arsenate (AsO_4^{3-})—which can be mistakenly joined to ADP by the phosphorylation enzymes—is highly toxic (Section 8).

Energy content of cellular ATP

$\Delta G^{0'}$ for ATP hydrolysis (under standard conditions) is –30 kJ/mol (above). In the cell, however, the ATP/ADP ratio is 100–1000, rather than 1 as under standard conditions, and ΔG for hydrolysis (ΔG_P) is around –55 kJ/mol.

Chemical features of ATP

There are other chemical (rather than energetic) features of ATP that may favour its use *in vivo*.
1 *Phosphate groups* are widely used to increase polarity (e.g. of glucose) in biological systems; ATP is a good source of such groups.
2 As an acid anhydride, ATP is a *good dehydrating reagent* for macromolecular biosyntheses, which typically proceed via condensation reactions.

3 As a triphosphate, ATP has *two modes of hydrolysis*; the occasional reaction requiring a driving energy of more than 50 kJ/mol can remove the two terminal phosphate groups.

4 *The adenine group* is frequently used as a 'tag' in coenzymes (NAD$^+$, FAD, coenzyme A, etc.); it plays no direct part in the reactions, but may provide a *favourable recognition site for enzymes*.

Transmembrane ion gradients

Transmembrane ion gradients, if they are not at equilibrium, can also serve as an energy store (Section 5). They are somewhat less flexible than ATP, being restricted to a particular location. Thus ion gradient energy is often *transduced* into the ATP disequilibrium before use—it is an *intermediate* between energy-releasing, redox reactions and the formation of ATP (see Section 1). There are, however, other (local) uses of gradient energy, notably in driving transmembrane transport or the bacterial flagellar motor. These applications are dealt with later (Sections 38, 39).

Storing energy in ion gradients

Energy is stored in an ion gradient (as in all other systems) by virtue of its displacement from equilibrium, i.e. because the electrochemical potential of the ion on one side of the membrane (*A*) differs from that on the other (*B*). Since, for an ion *X* (with a charge *z*)

$$\mu_A = \tilde{\mu}_A^{0'} + RT\ln[X]_A + zF\psi_A \qquad \text{(eqn 5.4)}$$

the difference in potential of *X* between sides A and B is given by

$$\Delta G_{A\to B} = \tilde{\mu}_B - \tilde{\mu}_A = RT\ln[X]_B/[X]_A + zF(\psi_B - \psi_A) \qquad \text{(eqn 7.1)}$$

i.e.   total free energy   concentration   electrical
       change   component   component

If the concentration and electrical components of $\Delta G_{A\to B}$ are exactly equal and opposite (see Fig. 5.2), the system is at equilibrium and *no energy is stored or needed to maintain it*.

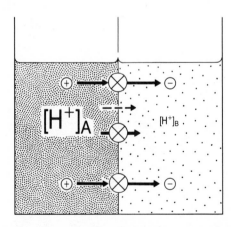

Fig. 7.4. Non-equilibrium distribution of H$^+$ in coupled membrane. The 'wheel' represents a coupling device driven by H$^+$ movement down its electrochemical gradient (arrows). The dotted arrow represents an uncoupled 'leak' of H$^+$. Compare with Figs 5.2 & 5.4.

If, however, the two components are complementary (see Fig. 7.4), energy is stored. From equation 7.1, we see that

1 for a tenfold concentration difference, ΔG becomes more negative by 5.9 kJ/mol, and

2 for a 100 mV voltage difference, ΔG becomes more negative by 10.3 kJ/mol (for a univalent ion).

These values are independent of the chemical nature of the ion (μ^0 is unchanged if an ion simply moves), although the electrical component is dependent on its charge, *z*. As an intermediate in ATP formation, H$^+$ ions are preferred as the gradient former in biological systems; evidence for this is given in Section 20. The reason why H$^+$ ions (rather than any other ionic species) are used probably lies in their direct involvement in the (energy yielding) redox reactions (Section 6).

Quantitative aspects of energy stored in a proton gradient

Measurement of ΔG for H$^+$ transfer ($\Delta\tilde{\mu}_{H^+}$) requires the separate assessment of pH inside and outside an organelle or bacterium, *and* measurement of the transmembrane electrical potential. Neither is simple, and measurement is discussed later in Section 23. However, various points emerge.

1 $\Delta\tilde{\mu}_{H^+}$ typically lies in the range 17–25 kJ/mol H$^+$.*

2 Comparing this value to ΔG for ATP hydrolysis (–50 to –55 kJ/mol), this means that more than one mole of protons is required to drive the synthesis of 1 mole ATP.

3 The distribution of $\Delta\tilde{\mu}_{H^+}$ between concentration and charge gradients varies widely between systems. For the chloroplast (thylakoids), a transmembrane ΔpH of 3.5 units provides $\Delta\tilde{\mu}_{H^+}$ of 21 kJ/mol H$^+$, and the charge gradient is negligible; in mitochondria, transmembrane electrical potentials of 150–200 mV ($\Delta\tilde{\mu}_{H^+}$ = 15–20 kJ/mol H$^+$) constitute most of the stored energy.

4 The amount of energy stored in a gradient, as in a chemical reaction, is potentially infinite—it will increase as the system moves further and further from equilibrium. The observed maximum, of about 25 kJ/mol, seems to be due to structural factors, such as the instability of membrane proteins to very low or very high pH (limiting the size of the pH gradient), and the dielectric breakdown of the membrane (limiting the size of the voltage gradient).

*$\Delta\tilde{\mu}_{H^+}$ is sometimes given in mV, rather than kJ/mol (100 mV = 10.3 kJ/mol). While this allows direct comparison with redox potentials, also measured in mV, the essential nature of $\Delta\tilde{\mu}_{H^+}$, as a *potential energy parameter*, must not be forgotten.

8 Substrate level phosphorylations

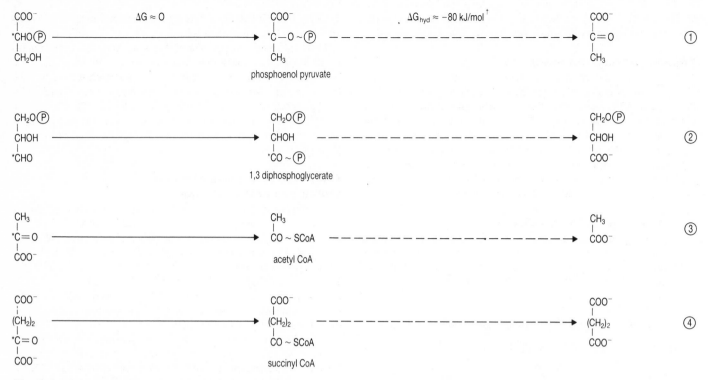

Fig. 8.1. Free intermediates in substrate level phosphorylation.
* Indicates carbon atom oxidized.
† Note that the hydrolysis reaction does not occur under normal circumstances; ΔG_{hyd} indicates the energy released if ATP were *not* synthesized.

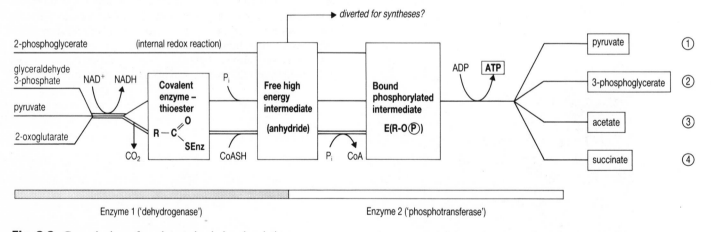

Fig. 8.2. General scheme for substrate level phosphorylation.

ATP synthesis involving chemical intermediates

Within living cells, certain reactions have a sufficiently negative ΔG (-50 to -80 kJ/mol) for one mol ATP to be formed per mol reactant used—if a *mechanism of coupling* exists. For reactions 1–4 in Fig. 8.1 such a mechanism does exist. Note that all these reactions are oxidations; in the final three, NADH is generated (the redox potential of the aldehyde or ketone is at least 500 mV more negative than that of NADH), while in the first there is an internal oxidation/reduction at C_2/C_3.

Using these reactions, the cell can make ATP without oxygen (or light). Reactions 1 and 2 are the ATP yielding steps in the familiar glycolytic fermentation of glucose to lactate, and reactions 3 and 4 occur in the tricarboxylic cycle. Clearly, however, for continued

ATP production via these reactions, NAD^+ must be regenerated from NADH (see below).

Chemical coupling mechanisms

Oxidation of glyceraldehyde-3-phosphate to 3-phosphoglyceric acid (by NAD^+) is a downhill process, energetically capable of ATP synthesis. *Coupling* occurs during the oxidation *if and only if* ATP synthesis occurs concurrently; the stages in ATP synthesis must somehow limit the progress of the oxidation. This is achieved by splitting up the oxidation into several steps which *share common intermediates* with ATP synthesis. In the case of the GAPdH reaction, oxidation of the aldehyde occurs early on—but the acyl group is not generated free until it has introduced first a phosphate and then an ADP into the overall reaction. In the pyruvate kinase reaction, phosphate is introduced before the redox reaction (at an earlier stage in glycolysis); after this the pyruvate is trapped as the enol phosphate until ADP enters the sequence (see Fig. 8.2).

Intermediate and dehydration reactions

Formation of the phosphoric acid anhydride, ATP, requires a dehydration. In the coupling mechanisms above, the dehydration occurs prior to phosphate anhydride formation. In reactions 2–4, the oxidation itself is linked to a dehydration and a *thioester* is formed. Sulphydryl (–SH) groups are convenient in this process because, apart from participating in the formation of a (thermodynamically unstable) thioester, since they also take part in redox reactions. In the GAPdH reaction, the participating –SH comes from a cysteine residue on the protein. In the other two reactions, it comes from the cofactor, lipoic acid.

Further stages in these reactions involve the generation of a mixed acid anhydride (acyl phosphate), and the required phosphoric anhydride (ATP), with eventual release of the free acid.

Water and the stability of intermediates

Since the intermediates in these pathways are dehydrated (have a high negative ΔG for hydrolysis in 55 M water), for the coupling pathway to be followed they, like ATP, must have *kinetic stability in water*. In principle, the enzyme's active site might prevent the access of water to the intermediate—but this does not appear to be the case here. Both the phosphorylated intermediates (in reactions 1 and 2) and the thioester intermediates (reactions 3, 4) can be released into solution and, indeed, may participate in other reactions. For example, acetyl CoA is widely used in biosyntheses, and 1,3-diphosphoglycerate may be used to produce 2,3-diphosphoglycerate.

Limitations of substrate level phosphorylation

1 For continued ATP production by the above reactions, NAD^+, which is present in limited amounts, must be regenerated from NADH. This is accomplished by transferring reducing equivalents to inorganic molecules like O_2 (in oxidative metabolism), CO_2, NO_3^-, SO_4^{2-} (in chemosynthetic bacteria) or to organic molecules (pyruvate, acetaldehyde) in fermentations. The higher the redox potential of the acceptor, the more energy is liberated; reduction of pyruvate yields insufficient energy to form ATP, but reduction of NO_3^- or O_2 respectively yields enough energy to form 1 or 3 mol ATP/mol NADH. Without additional coupling mechanisms, this energy would be wasted.

2 Coupling is 'efficient' (formation of ATP occurs with little net loss of free energy as heat) *only if $\Delta \overline{G}$ (kJ/mol reactant) matches $\Delta \overline{G}$ for ATP formation*. It is difficult to see how a one step oxidation (of 1 mol NADH, say) could be coupled directly to formation of 3 mol ATP, or, conversely, how oxidation of two molecules could co-operate in one phosphorylation, if generation of some anhydride intermediate(s) were required.

3 *A different set of enzymes is required for every coupled reaction.* Although there are similarities between the mechanisms of reactions 2–4, each of the 4 reactions requires a pair of (fairly complex) enzymes different from the others. Clearly, if each ATP generating reaction followed such a path, a large number of enzymes would be required.

These last two problems are overcome by using, instead of a unique 'high energy' intermediate for each oxidation, *a single common intermediate between the oxidation of a variety of substrates and ATP formation*. This intermediate is a transmembrane proton gradient. Understanding this (more convenient) mechanism of ATP formation occupies much of the rest of this book.

9 Free oxygen: its benefits and dangers

Oxygen and the atmosphere

At its simplest, living matter may be said to comprise, largely, carbon in various states of reduction. Reduced carbon compounds are not at equilibrium with an oxygen-containing environment—hence their oxidation (to $CO_2 + H_2O$) will yield energy. Indeed, our energy derives from the oxidation of food, using atmospheric dioxygen (O_2) as the final electron acceptor.

This idea is so familiar that it is easy to forget that, when life appeared, oxygen existed only in combination (as water, carbonates etc.). The atmosphere itself contained the reducing agents CO, H_2 (as well as H_2O, CO_2 and N_2) and could not oxidize organic compounds with the release of energy. Indeed, some energy was available from reducing carbon compounds (e.g. $CO_2 \rightarrow CH_4$); the rest came from the sun.

Dioxygen appeared in the atmosphere from the photolysis of water. This process was a biological one, oxygen-evolving photosynthesis, which evolved with the cyanobacteria some 2×10^9 years ago. It is still a major energy-trapping process today, in the cyanobacteria and higher plants, and is described in detail in Section 15.

The appearance of atmospheric oxygen altered the environment of living organisms, and profoundly influenced their nature. On the positive side, the presence of a strong oxidizing agent in the environment allowed the evolution of oxidative energy production, and the evolution of heterotrophic life styles. Initially, however, the effects were deleterious. Oxygen could oxidize organic structures and lead to the destruction of organized cell components. There were two options—to hide (as in the anaerobes) or to develop protective mechanisms to limit oxygen damage. All aerobic organisms today contain such mechanisms; some are discussed below. Nonetheless, it is interesting to think of the cyanobacteria as the first major polluters of the Earth's atmosphere, introducing toxic free oxygen. Luckily, in the long term, the benefits outweighed the disadvantages.

Oxygen is a good oxidizing agent

Oxidation of glucose by oxygen, to $CO_2 + H_2O$, has $\Delta G^{0'} = -2880$ kJ/mol glucose; ΔG *in vivo* is even more negative due to the low concentration of product CO_2. Comparing this to $\Delta G^{0'} = -200$ kJ/mol for anaerobic glycolysis (to lactate), we see that oxygen greatly increases the energy available to the organism from each glucose molecule.

This is reflected in the very positive redox potential of dioxygen—Table 9.1 shows it to be a very good oxidizing agent relative to other

electron acceptors in the environment (NO_3^-, CO_2, SO_4^{2-}), and to organic oxidants such as NAD^+ or quinones. Thus we find *in vivo* that some organisms (like *E. coli*) can use alternative electron acceptors to oxygen—but they will generally switch to oxygen if it becomes available, to maximize their energy production.

Reactions of dioxygen

The pathway of O_2 reduction is not simple. Addition of 1 e^- yields the unstable species, superoxide (O_2^-). The instability of superoxide ($E^{0'} = -0.33$ V) is such that, kinetically a pathway of O_2 reduction involving *two electron steps* is preferred (Fig. 9.1); cytochrome oxidase and most other oxidases follow this principle.

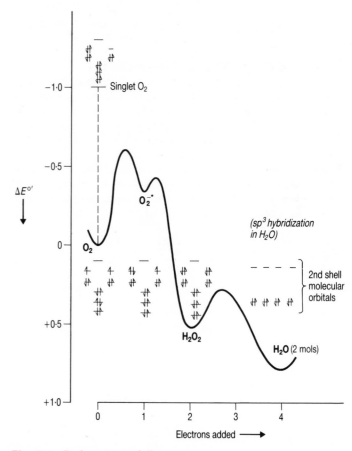

Fig. 9.1. Redox states of dioxygen.

There are further problems for biological systems. Addition of 2e^- to oxygen yields hydrogen peroxide. *The chemical reactivities of O_2^- and H_2O_2 (and the related species, hydroxyl radical, OH) are very great* and they will tend to damage organic compounds. Lipids, particularly those with unsaturated fatty acids, will be oxidized, and membranes and other structures will be damaged.

The cell uses three strategies to minimize these effects.

1 Only a *very limited number* of reductive enzymes bind O_2, and in these the intermediates are *bound very tightly*.

2 The cell contains *reductants* such as **ascorbic acid** and **glutathione** (soluble), and **carotenoids** and **vitamin E** (membrane bound), which preferentially react with the small fraction of these powerful oxidants that are released from enzymes during O_2 reduction.

3 The cell contains *degradative enzymes*, such as **superoxide dismutase** and **catalase**, which rapidly destroy superoxide and H_2O_2.

Table 9.1. Oxygen and alternative electron acceptors in biological reductions

Redox couple	$E^{0'}$	$\Delta G^{0'}$ for reduction by NADH at pH 7.0
$NAD^+/NADH$	-0.32 V	
$2H^+/H_2$	-0.42 V	$+19.3$ kJ/mol NADH*
SO_4^{2-}/S^{2-}	-0.20 V	-23.1 kJ/mol NADH
CO_2/CH_4	$+0.17$ V	-94 kJ/mol NADH†
NO_3^-/NO_2^-	$+0.42$ V	-142 kJ/mol NADH
$\frac{1}{2}O_2/H_2O$	$+0.82$ V	-220 kJ/mol NADH

* ΔG will be more negative *in vivo*, due to the insolubility of the product.
† Reductant *in vivo* is H_2. ΔG will be more positive *in vivo*, as H_2 is present at very low concentration. At 10^{-4} Bar H_2, $\Delta G = -25$ kJ/mol.

Dioxygen is a kinetically stable molecule

Organic molecules do not spontaneously combust in oxygen; despite the large, negative ΔG, oxidation of glucose by air proceeds at a negligible rate unless catalysts are present. Thus glucose in oxygen, like ATP in water, fulfils the two requirements for an energy store—it is *thermodynamically unstable but kinetically stable*.

Dioxygen, indeed, is a relatively unreactive molecule. Mixing $H_2 + O_2$ at room temperature gives no observable change, while mixing $H_2 + Cl_2$ gives an explosive reaction. This relates to its electronic structure; in O_2 10 of the 12 electrons in the valence shell pair in (bonding) molecular orbitals but the other two *remain unpaired with identical spins* (triplet oxygen; paramagnetic)—see Fig. 9.1. On thermodynamic grounds, a 1 e^- reduction is unlikely (above); for a 2 e^- reduction, 2 electrons with the same spin must be added to the π^* orbitals. This is difficult because an added pair of electrons would normally come from a single donor orbital, and thus have opposing spins. Thus dioxygen is kinetically 'protected' against addition of two electrons (e.g. from an organic donor).

Reduction of O_2 becomes possible in the presence of a *paramagnetic catalyst* (e.g. haem iron, Cu^{II}), which 'scrambles' the spins from the donor. Reduction of dioxygen by cytochrome oxidase is discussed in detail in Section 13.

Superoxide dismutase

All aerobic cells contain superoxide dismutases, catalysing the reaction

$$O_2^- \cdot + O_2^- \cdot + 2H^+ \rightarrow H_2O_2 + O_2$$

From Fig. 9.1, superoxide is a fairly good reducing agent ($E^{0'} = -0.33$ V) as well as a good oxidizing agent—hence the *dismutation* where one ion is oxidized by another identical one. Superoxide dismutases are very efficient enzymes and destroy each O_2^- they encounter (diffusion limited). Thus the deleterious effects of superoxide are prevented.

There are two families of superoxide dismutase, the copper–zinc enzymes (largely of eukaryotic origin) and the manganese or iron enzymes (found in prokaryotes and organelles). The similarity of the prokaryotic and organelle dismutases may relate to the prokaryotic origin of organelles (Section 45). Amino acid sequences are homologous within a family, but unrelated between families.

Superoxide dismutases work by a two step mechanism—the metal centre is first reduced by O_2^-, releasing O_2 and then reoxidized by the second O_2^-, producing H_2O_2.

Partially reduced oxygen as a toxic agent

The damaging effect of partially reduced dioxygen can be exploited as a killing mechanism. This occurs both at the *cellular level*, and at the level of *conflict between organisms*.

1 The function of polymorphonuclear lymphocytes is to kill invading bacteria. This is achieved using partially reduced oxygen (H_2O_2) and **myeloperoxidase** to generate a series of highly reactive species (including singlet oxygen, see Fig. 9.1) which when released close to the invading bacterium, seriously damage it. Note that HOCl, hypochlorous acid, is a major component of commercial bleach.

$$O_2 \xrightarrow{2H^+ + 2e^-} H_2O_2 \xrightarrow[\text{Myeloperoxidase}]{2H^+ + 2Cl^-} 2HOCl \rightarrow 2HOOCl$$
$$\rightarrow 2HCl + 2O_2$$

2 Herbicides are used by man to kill unwanted plants. **Paraquat** is a redox agent which acts by diverting highly reducing electrons, generated in photosynthesis, to O_2. This produces large quantities of O_2^-, which swamp the plants' superoxide dismutase activity, leading to damage to plant tissues (see Section 16).

10 Cofactors in oxidation/reduction reactions

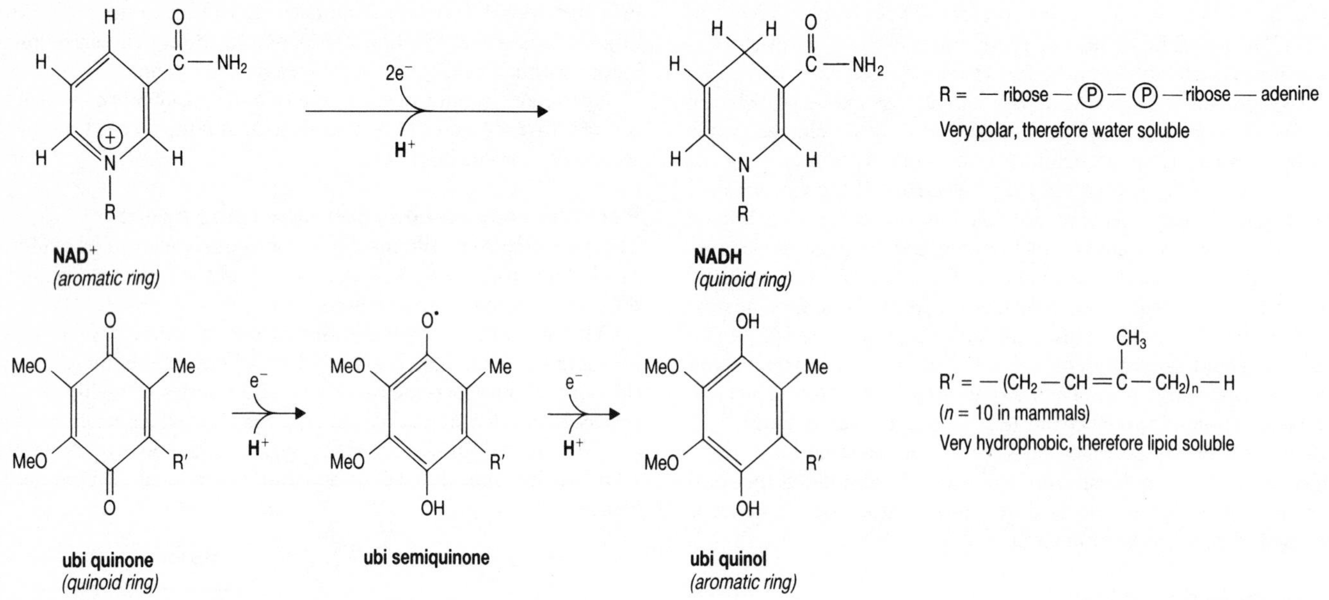

NAD⁺
(aromatic ring)

$2e^-$ ⟶ H^+

NADH
(quinoid ring)

R = — ribose — Ⓟ — Ⓟ — ribose — adenine
Very polar, therefore water soluble

ubi quinone
(quinoid ring)

e^- / H^+

ubi semiquinone

e^- / H^+

ubi quinol
(aromatic ring)

$R' = -(CH_2-CH=\overset{\overset{\displaystyle CH_3}{|}}{C}-CH_2)_n-H$
(*n* = 10 in mammals)
Very hydrophobic, therefore lipid soluble

Fig. 10.1 Organic cofactors of redox reactions.

Organic and inorganic cofactors

Proteins involved in redox reactions must accept and donate electrons. However, the constituents of proteins, the amino acids, cannot accept electrons; they are not easily reduced or oxidized. Thus, in biological systems, electrons are carried by *cofactors* of the proteins. These may be organic (**nicotinamide**, **flavins** and **quinones**) or inorganic (Fe^{II}, Cu^{II}, Mn^{II}). Other cofactors (e.g. lipoic acid, Ni^{II}) are also involved in some specific classes of redox reaction (Sections 8 & 14).

Organic cofactors

The oxidized and reduced forms of NAD and quinones are given in Fig. 10.1. In one form, the cofactor has *aromatic* character; in the other, it is conjugated but not aromatic (*quinoid* structure). These compounds (together with flavins and O_2) share the property that on reduction, H^+ *must be taken up to neutralize the negative charge due to the added electrons* (see Section 6).

Nicotinamide, flavin and quinone cofactors differ in $E^{0'}$, NAD(P)H being the best reducing agent (most negative $E^{0'}$) and OH₂ the worst. They also differ in **mobility**. NADH can *diffuse in aqueous solution* (e.g. the mitochondrial matrix), allowing it to shuttle electrons between soluble and membrane-bound enzymes. Quinones are lipid soluble and can *diffuse* rapidly *within the membrane*, and so shuttle electrons between intrinsic membrane proteins. Flavins are each *bound* permanently to one specific enzyme (e.g. succinate dehydrogenase).

All these cofactors undergo a 2 e^- reduction. This may, in principle, involve two 1 e^- steps, the first producing an unstable 'semiquinoid' structure (Fig. 10.1). The inherent instability of this structure (like O_2^-, it will dismutate into oxidized and reduced forms) means that simultaneous addition of two electrons is preferred mechanistically. However, the unstable species may be stabilized *in vivo* by its binding to a protein and, at least in the case of the quinones, a **semiquinone** may exist long enough within a membrane to play a significant role in electron transfer (Section 24).

Fe centres in proteins

Like the flavin cofactors, Fe participating in redox reactions is tightly bound to proteins (as in Cu); it cannot diffuse through the aqueous solution nor through the lipid membrane. Fe as a redox cofactor can exist in one of two basic structures.

1. Iron–sulphur proteins

These contain Fe liganded to **inorganic sulphide** (and to **cysteine–SH**) within the protein. The Fe atoms are essentially tetrahedral. Many such proteins contain $4Fe/4S^{2-}$ clusters arranged in a distorted cube (Fig. 10.2) with three of the 4 cys ligands close together in the protein sequence (. . . cys-xx-cys-xx-cys). Although each cluster contains 4 Fe atoms, it is stable only in a *mixed valence state* (at least 1 Fe^{II} and 1 Fe^{III} present). Thus only three oxidation states are possible

$$Fe^{II}.3(Fe^{III}) \xrightarrow{e^-} 2(Fe^{II}).2(Fe^{III}) \xrightarrow{e^-} 3(Fe^{II}).Fe^{III}$$

State (1)   (0)   (−1)

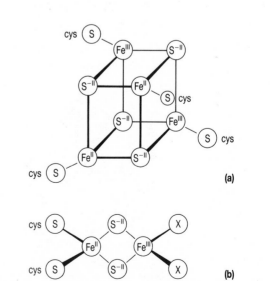

Fig. 10.2. (a) 4Fe4S centre, e.g. succinate dehydrogenase.
(b) 2Fe2S centre: X is S-cys in low $E^{0'}$ proteins (e.g. ferrodoxin); X is N-his in high $E^{0'}$ proteins (Rieske FeS protein).

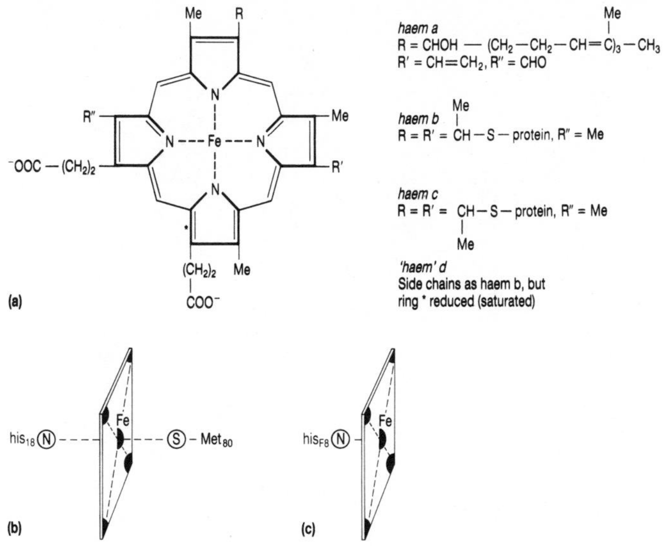

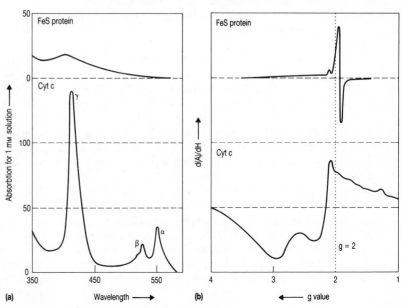

Fig. 10.3. Haem group of cytochromes.
(a) Face view, showing Fe coordination within haem ring.
(b) Side view, showing coordination from protein in cyt c (e$^-$ carrier).
(c) Side view, showing coordination from protein in haemoglobin (O$_2$ carrier).

In State (-1), the added electron is shared between all atoms in the cube; it is readily lost and thus this transition has a negative $E^{0'}$. Removing an electron from State (0) is more difficult: $E^{0'}$ (1/0) is positive.

While all three oxidation states are available in principle, in any one protein only two are allowed. Thus 4Fe4S proteins fall into two classes: (i) those which are *strong reducing agents* ($E^{0'} < -0.2$ V, State -1 to 0) such as those at the reducing end of the mitochondrial electron transfer chain, and (ii) those which are not ($E^{0'} > 0.0$ V, State 0 to 1), as are found in some photosynthetic bacteria. The latter are known as the *high potential iron proteins*, HiPiP.

Some iron–sulphur proteins contain 2Fe2S clusters (Fig. 10.2). These can similarly change between the Fe^{II}/Fe^{III} and either the 2 Fe^{III} or 2 Fe^{II} states. One, the Rieske iron–sulphur protein, is found at the oxidizing end of the mitochondrial electron transfer chain and several are found at the reducing end (NADH dehydrogenase) (Section 11).

Because of their inorganic sulphide content and their sensitivity to oxygen, it is thought that iron–sulphur proteins represent primitive electron carriers, having evolved before free oxygen appeared. This view is supported by experiments in which 'model' FeS centres can be created simply by heating FeS, Na$_2$S and an organic thiol.

2. Cytochromes

Cytochromes contain Fe bound to 4 nitrogen atoms in a **haem ring**, and to N or S ligands from the protein. The Fe atoms are *octahedral*. There are four types of haem group found in the cytochromes, denoted *a*, *b*, *c*, *d*, which differ in the side chains attached to the ring (Fig. 10.3a). (For historical reasons, cytochromes denoted *f* and *o* are found in the literature; these fit into classes *c* and *b* respectively.)

Within a class, the fifth and sixth ligands to the iron may vary. Histidine commonly occupies one position; the other may also be histidine, or methionine (sulphur) or, indeed, water (i.e. no fixed ligand) (Fig. 10.3b). If there is no fixed sixth ligand, oxygen (or H$_2$O$_2$) can ligand to Fe and the cytochrome can reduce it; it becomes an **oxidase** (or peroxidase). If all six ligand positions are filled, O$_2$ cannot bind and the protein is simply an **electron carrier**. Oxidases occur within all four classes of cytochrome.

Spectroscopy of Fe centres in proteins

Since most Fe proteins within cells are membrane bound (and hence difficult to purify), spectroscopic methods are used widely in their study. In cytochrome, the haem iron imparts a bright red colour, with clear absorption bands in the visible region (Fig. 10.4a). The spectrum changes with oxidation state, allowing this to be monitored (see Section 6). The reduced species gives the sharpest bands, the position of the band at highest wavelength (α band) frequently being used to differentiate cytochromes within one class—hence cytochromes b$_{562}$, b$_{566}$ in mitochondria, cytochromes c$_{550}$, c$_{553}$ in photosynthetic bacteria.

FeS proteins are not coloured—although they absorb light weakly in the visible range (Fig. 10.4a), no band stands our clearly. Visible spectroscopy of FeS proteins is thus difficult, particularly in membrane preparations.

Both cytochromes and FeS proteins contain **unpaired electrons**, and hence give absorption bands in **electron paramagnetic resonance** (EPR) spectroscopy (Fig. 10.4b). Oxidized cytochromes (Fe^{III}) give major absorption bands at *g* values around 2 or 2.2 and 2.8—depending on the spin state of the iron, while reduced FeS proteins (State -1, above) have a very characteristic peak at $g = 1.9$–1.96. The surprisingly low value of *g* in reduced FeS proteins probably relates to the delocalization of electrons between the iron atoms.

Fig. 10.4. **(a)** Visible spectra of cyt *c*.
(b) EPR spectra of cytochrome and iron–sulphur proteins.

11 The mitochondrial respiratory chain: 1 Path of electron flow

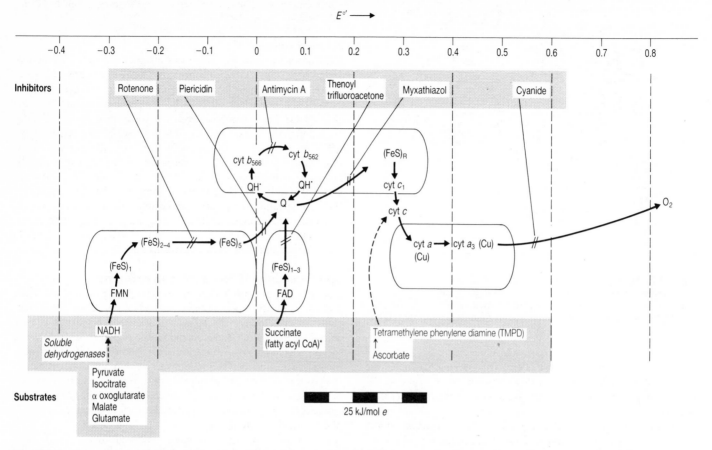

Fig. 11.1. The respiratory chain in animal mitochondria. Components associated within a single complex of polypeptides are outlined. * Fatty acyl CoA dehydrogenase is a complex distinct fromm succinate dehydrogenase, but which also has $E^{0'} \approx 0$. There are five different FeS centres in NADH dehydrogenase (FeS_{1-5}), three in succinate dehydrogenase and one associated with cyt c_1 (FeS_R), the Rieske FeS protein.

Electron transfer occurs between specific pairs of redox carriers

Oxidation of a good reducing agent, like NADH, by a good oxidizing agent, like O_2, yields considerable energy. At levels of NADH and O_2 in the cell ($[NADH]/[NAD^+] = 10$, $[O_2] \approx 0.23$ atmos.), $\Delta G = -225$ kJ/mol NADH, just slightly more negative than $\Delta G^{0'}$. This energy is not released in a single reaction but in a series of steps, each yielding some convenient 'quantum' of energy (about 25 kJ/mol; see Section 5).

Energy release is achieved by the progression of electrons through a *series of carriers of increasing redox potential*—a **respiratory chain** (Fig. 11.1). The carriers are (generally protein bound) cofactors of the types shown in Section 10. The approximate redox span required to yield 25 kJ/mol e^- transferred is indicated on the figure. To ensure that the flow is ordered, the electron must pass from one specific carrier to another. Transfer between the wrong pair of carriers would release too much energy for the conservation mechanism, and must be stopped.

Random electron transfer is prevented (i) by the existence on the surface of an electron transfer protein of a *binding site specific for its correct partner*, and (ii) by *restricting the free diffusion* of the cofactors so that most are unable to collide with each other. Specific aspects of protein structure and membrane organization are dealt with in Sections 12 and 13. This section considers the sequence of carriers itself—the pathway for an electron from (say) NADH to O_2.

Factors affecting the redox potential of Fe

The position of an electron carrier protein in the respiratory chain depends, to some extent, on its cofactor. *4Fe4S centres*, converting between States (0) and (−1) have a highly negative redox potential (Section 10). They are found at the *reducing end* of the respiratory chain. *Haem cofactors* have a more positive redox potential—they are found in proteins near the *oxidizing end*. However even proteins with identical cofactors may vary considerably in $E^{0'}$. In nature, c-type cytochromes have been found with $E^{0'}$ from $-250 \rightarrow +390$ mV (although the two in mitochondria differ only by about 20 mV). These all have the same haem ligand; thus variations in $E^{0'}$ must also involve the surrounding protein.

In principle, *any feature of the protein which stabilizes the oxidized form of the cofactor, will make $E^{0'}$ more negative*, i.e. make it a better reducing agent. In extremis, one oxidation state may be lost completely; in mitochondrial 4Fe4S proteins, removal of an electron from $3Fe^{II}Fe^{III}$ (State (−1)) is permitted but State (0) is so stable compared with State (+1), that a second electron is never lost (Section 10).

Factors which affect the redox potentials of cofactors in proteins are:

1 *Type of ligand bound to the metal.* In haem proteins, ligands 5 and 6 may be his/his (as in mitochondrial cyt b), his/met (as in mitochondrial cyt c) or possibly some other nucleophiles. Due to ligand field effects on electron distribution, and on stability of the spin states of iron, Fe^{II} and Fe^{III} will be stabilized differently; in particular, a more basic ligand (e.g. histidine) will increase electron density around Fe making it a better reductant ($E^{0'}$ decreased).

2 *Structural constraints within the protein.* Oxidation of Fe^{II}—and changes in spin state—will alter the radius of the iron atom. If the surrounding protein is fairly rigid, one form will be preferred; in the other, the bonds to the iron will be 'strained' (the Fe will be in its 'entatic' state).

3 *Charge on the protein.* Oxidation of a metal leads to an increase in positive charge—thus it is favoured in a protein of overall negative charge. If 2 extra COO^- groups are engineered (genetically) onto the surface of cytochrome *c*, its redox potential decreases (by up to 100 mV). Thus charge near the iron or even on the protein surface may affect the redox potential of a cofactor.

Oxygen binding

Random protein–protein electron transfer is prevented in the respiratory chain, as noted above. Similarly, electrons on a carrier protein must be prevented from reaction with the strong oxidizing agent, O_2. This is achieved by *preventing access of oxygen to the redox metal.* FeS proteins have all 4 available sites on Fe liganded with S, and the cytochromes typically have all 6 sites liganded (4 haem N and 2 protein ligands); thus there is no space available for oxygen binding. In contrast, where oxygen is involved in an oxidation (e.g. of mitochondrial cytochrome a_3), the haem Fe bears only one ligand from the protein, the other ligand being H_2O (when reduced) or O_2 in the oxygenated state (see Fig. 10.3).

Determining the order of respiratory chain components

1. Standard redox potentials

Electrons will flow from carrier *A* to carrier *B* when the redox potential of *B* is the higher ($\Delta G'$–ve). Assuming that *E* is close to $E^{0'}$ carriers should be ordered in the chain by increasing $E^{0'}$ values. This approach places flavins and FeS proteins close to NADH (reducing end of the chain) and cytochromes (and Cu) closer to O_2 (oxidizing end).

This approach assumes:

1 $E^{0'}$ can be measured accurately; not always easy for membrane proteins (Section 6).

2 $E \approx E^{0'}$. This will not be the case if [ox]/[red] ratio *in vivo* is not close to one, or if the pH that the carrier experiences (not necessarily the cytoplasmic pH) is not 7 (see Section 6). These effects can be significant—a tenfold departure from standard concentrations ([ox]/[red] = 10 or pH = 6.0) leads to a shift in *E* of $RT/F \ln 10 = 60$ mV.

2. Inhibitors of electron flow/alternative substrates

In the presence of an electron donor + oxygen, an inhibitor of electron flow (binding to one of the electron transfer proteins) will block the chain, leading to a reduction of all components on the reducing side of the block, and oxidation on the oxidizing side (this can be determined spectroscopically). A series of inhibitors can thus be used to delineate the chain.

The blocks can be circumvented by alternative reductants or oxidants—NADH oxidation, but not succinate oxidation, is blocked by rotenone—and a picture of the chain is build up (see Fig. 11.1).

3. Rapid reaction kinetics

Anaerobic mitochondria are given a pulse of oxygen, using a rapid mixing system, and the progress of oxidation followed. Cytochrome aa_3 is oxidized within 2 ms, cyt *b* within 20 ms and so on.

Respiratory chain model

The resulting model of the respiratory chain in *animal* mitochondria is shown in Fig. 11.1. In outline, there are several membrane bound **dehydrogenases**, one specific for NADH (produced by soluble dehydrogenases) and others specific for other substrates (succinate, fatty acyl CoA, glycerol 3 phosphate). All contain 4Fe4S centres and flavin. The electrons then reduce a mobile **pool of quinone**, whence they follow a common pathway through several cytochromes (and a 2Fe2S protein) to an **oxidase** and finally *oxygen*. (In bacteria, where alternative electron acceptors may be used, the chain may branch again after the quinone pool, see Section 14.)

For many years, it was tacitly assumed that this 'chain' of electron carriers was linear, i.e. an electron passed through each carrier in turn. Anomalies in the kinetics of oxidation, and of behaviour of inhibitors (especially antimycin) suggest that this may not be absolutely true, as indicated in the figure. In particular, electrons passing through the *b* cytochromes are not used directly to reduce the 2 Fe2S centre. This is discussed further in Section 24.

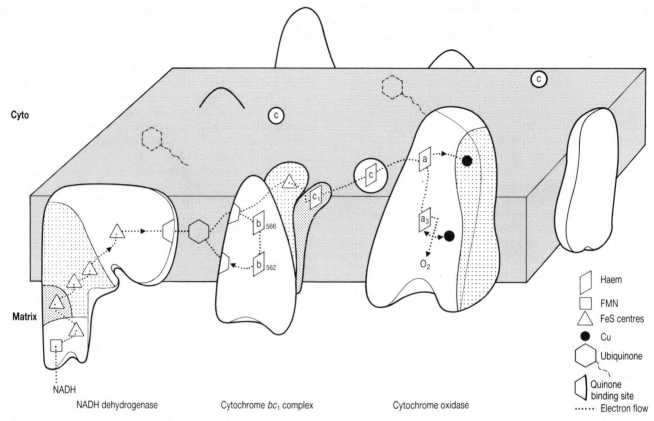

Fig. 12.1. Topology of electron transfer chain in mitochondria. Only polypeptides bearing cofactors are distinguished.

In the mitochondrial membrane, electrons pass from one redox cofactor to the next, in sequence (Section 11). The membrane contains about 70% protein (the remainder being lipid) by mass, of which about half are involved in electron transfer.

These proteins bind the cofactors and determine:
1 their location within the membrane
2 their mobility
3 the electron donors and/or acceptors allowed access to a cofactor
4 the changes on protonation consequent on oxidation and reduction.

The **location** and **mobility** of electron transfer cofactors will be dealt with in this section, and the other two points in later sections.

Respiratory chain complexes

Some 60 different polypeptides are involved in the transfer of electrons from NADH to O_2. Only a few, therefore, can interact directly with redox cofactors. However, within the membrane, these 60 polypeptides are not free; those that do not themselves interact with cofactors associate with a specific subset of the interacting peptides such that nearly all the peptides can be assigned to one of three aggregates—the **respiratory complexes**.

Within the membrane, therefore, are assemblies (*complexes*) of strongly interacting peptides which function as integrated units (Fig. 12.1). Within such a complex the (non-covalent) interactions are sufficiently strong that the membrane lipids can be dissolved (using a mild detergent, like cholate), leaving the complexes still assembled, and suitable for purification.

The activities of the complexes span the respiratory chain. Thus we have:
● Complex I: *NADH: ubiquinone oxidoreductase*

● Complex III: *ubiquinol: cyt c oxidoreductase*
● Complex IV: *ferrocytochrome c oxidase* (which reduces oxygen)

Each complex comprises a small number of peptides that bind the cofactors, plus several others. In complex IV (cyt c oxidase), for example, the two largest subunits (I, II) bind the copper and haem, and there are some 11 other peptides.

One role of these additional peptides is to *restrict access of reductants and oxidants*; a three subunit protein derived from complex I containing flavin and two of the FeS centres will oxidize NADH but can transfer electrons to cyt *c* as well as the physiological acceptor, ubiquinone. Other functions of these additional peptides may be *regulation* (e.g. in tissue specificity, see Section 35), H^+ *pumping* (see Section 25) or in *assembly*.

Note that the mitochondrial membrane contains other polypeptide complexes in addition to the above. Examples include succinate dehydrogenase ('Complex II'), fatty acyl CoA dehydrogenase (Section 11), and the ATP synthase ('Complex V'; Section 25). Further examples are found in other membranes (see later).

Location of respiratory chain proteins in the membrane

Several polypeptides of the respiratory chain *traverse the lipid bilayer*, probably as one or more α helices. They are thus integral membrane proteins. Examples include subunit I of complex IV (which carries both haems *a*) and the 42 kDa subunit of complex III (carrying both haems *b*) (Fig. 12.2). In both cases, *the two haem groups lie on opposite sides of the membrane*, perpendicular to its surface.

Other polypeptides are restricted to one side of the membrane; these associate with specific binding sites on integral proteins.

Cytochrome c lies on a separate, peripheral protein on the cytoplasmic face (**C side**) of the membrane. Complex I (NADH hydrogenase) appears to have all six of its cofactors (FMN, $5 \times$ FeS centres) bound to peripheral membrane proteins on the matrix face (M side), the transmembrane region serving as a membrane anchor for these proteins (Fig. 12.1).

Most of the transmembrane peptides traverse the membrane several times—the 42 kDa subunit of cyt b crosses the membrane 8–9 times. Here, the presumed histidine ligands to the iron lie on helix 2 (his 85, his 99) and helix 4 (186, 201) (Fig. 12.2). In addition, a considerable fraction of the complexes lies outside the membrane—the bulk of complexes II and III lies in the mitochondrial matrix and of complex IV in the intermembrane space (Fig. 12.1).

Approaches to the topology of the mitochondrial membrane are outlined in Table 12.1. Electron diffraction studies, apart from giving overall shape, indicate that complexes III and IV can dimerize, at least in two-dimensional crystals. Whether or not they are dimeric in the membrane is uncertain.

Mobility of electron carriers

The mitochondrial mosaic is a fluid mosaic of lipid and proteins. Within it, the electron carriers can *diffuse laterally* (Fig. 12.1); the respiratory complexes at about 4×10^{-10} cm^2 s (leading to collisions 10–100 times per second) and *ubiquinone some 10–100 times faster*. Since the concentration of ubiquinone is about 10 mol/mol complex III, it is probable that electrons leave complex I (or II) by collision with ubiquinone which then diffuses rapidly to complex III. Unlike the protein complexes, ubiquinone (or ubiquinol) can also *flip across* the membrane, allowing it to react at binding sites facing the M side or the C side of the mitochondrion (Section 24).

Table 12.1. Approaches to topology of mitochondrial respiratory chain

Approach	Type of information
Substrate accessibility	NADH binding site on *M side**
Salt wash of membranes	Cyt c on *C side* Succinate dehydrogenase on *M side*
Detergent solubilization of membrane	Identification of complex assemblies
Accessibility of water-soluble reagents (e.g. antibodies, diazobenzene sulphonate, proteases)	Subunits I, II, III of cyt oxidase span the membrane
Accessibility of membrane-soluble reagents (e.g. arylazido phospholipid, iodonaphthylazide)	Several small subunits of NADH dehydrogenase (not bearing the cofactors) span the membrane
Cross linking reagents	Cyt c interacts with subunit II of cyt oxidase
Gene/protein sequences	Predict 8–9 transmembrane helices and two haem binding sites on 42 kDa protein of complex III
Electron diffraction in two-dimensional crystals	Bulk of complex III protrudes on *M side* Bulk of complex IV protrudes on *C side* Complex IV is forked. Dimers?

* The designations M (Matrix, inside the mitochondrion) and C (Cytoplasmic, outside the mitochondrion) give the orientations of the sides of the membrane relative to the rest of the cell (see text).

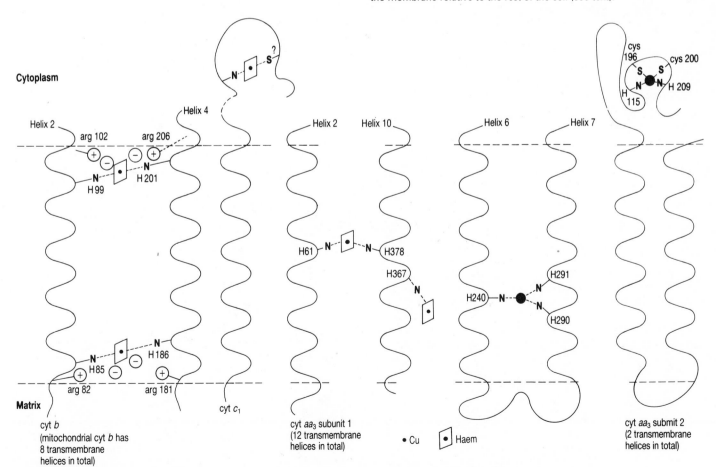

Fig. 12.2. Models for cofactor binding in cytochromes aa$_3$ and bc$_1$. The wavy line indicates a transmembrane helix; numbering is from the N terminus. H indicates a histidine residue liganded to metal. Note that the negative side chains of the b porphyrin interact with the positively charged arginine residues.

Cytochrome c is an unusual electron carrier in being water-soluble, dissolved in the fluid between the internal and external mitochondrial membranes. Within this space, it can diffuse freely and it seems likely that it *shuttles electrons between the relatively immobile* cytochromes c_1 (of complex III) and a (of complex IV). Since electrons enter and leave cytochrome c at the same face of the molecule (see Section 13), *rotation and diffusion of this molecule is needed for electron transfer.*

The rate of electron flow, in mitochondria, is limited by this (diffusive) transfer between complexes, i.e. by collision. If collision frequency is reduced, by artificially 'diluting' the membrane with added phospholipid, electron flow is reduced—and rates can be restored by adding more quinone. Within complexes, where cofactors are held rigid (in a presumably ideal configuration) electron transfer is much faster ($> 100\,s^{-1}$).

Quinone binding sites

Ubiquinol/ubiquinone can be looked on as a lipid soluble substrate for the oxidoreductase enzymes in the respiratory chain. Each of the complexes, I, II, III, and fatty acyl dehydrogenase, must have a quinone binding site, presumably in the lipid phase, with a distinct affinity. However, because lipids and detergents are present in preparations of these complexes, the location of these sites (on which polypeptide, for example) and their binding affinities are not known.

However, it is clear that quinone binding to one or more of these complexes *stabilizes the semiquinone radical*, i.e. the semiquinone binds better than either the fully oxidized or reduced form. This means that, in contrast to the situation with dissolved quinone (Section 10), reduction (or oxidation) can proceed in $1\,e$ steps and the quinone can, in fact, provide a **gate** between $2\,e$ reductions of the organic redox cofactors (NADH, $FADH_2$) and the inorganic cofactors (Fe, Cu; Section 10).

Complex III, indeed, bears two quinone binding sites, at different positions on the complex (Fig. 12.1). (Access of electrons to one site is sensitive to antimycin A, and to the other to 2,3 mercaptopropanol.) Both of these sites stabilize the semiquinone radical well, and both seem to participate in the 'branched chain' region of electron flow (Section 11). Orientation of the two sites differs with respect to the membrane surfaces; one (Q_P) is accessible from the C side of the membrane, and the other (Q_N) from the M side.

13 Electron transfer mechanisms

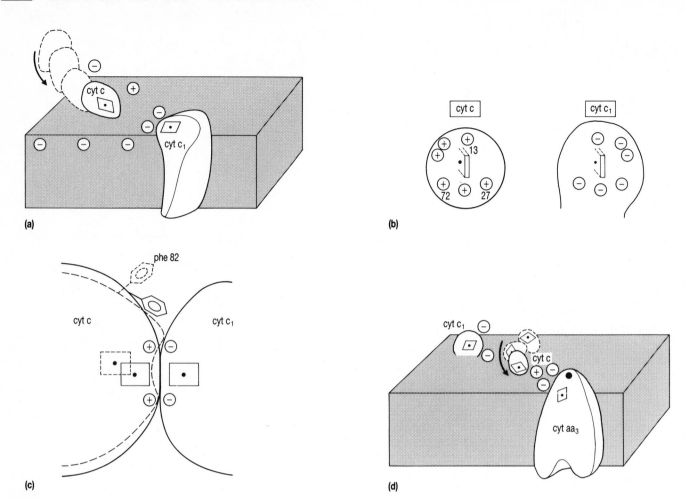

Fig. 13.1. Electron transfer between cyt c_1, cyt c and cytochrome oxidase.
   (a) *Orientation.*
   (b) *Docking.* Haem edge shown centrally. $\oplus$ = positive lys residue, $\ominus$ = negative carboxylate (gly/asp) residue.
   (c) *Relaxation to the functioning configuration.* Initial complex shown dotted, final complex shown in solid lines.
   (d) *Release and rotation.*

Inner and outer sphere mechanisms

The Fe atoms in cytochromes (except those binding oxygen) are coordinated in all six possible positions—four from the haem ring and two from the protein. It is impossible, therefore, for Fe atoms on two cytochromes to approach each other within bonding distance (3–5 Å). Electrons must be transferred through the environment of surrounding ligand—an **outer sphere** transfer.

Under these conditions, the rate of electron transfer between redox centres falls off exponentially with the distance between them. In other words, the protein apparently acts as a (passive) 'organic glass', and not as an electron conductor. Electrons 'hop' from centre to centre by quantum mechanical **tunnelling**; *the role of the protein is simply to position the redox centres at a suitable distance.*

Quantum mechanics allows entities as small as electrons to 'tunnel' over distances up to about 20 Å. Thus, for electron transfer, the structure of the cytochromes must allow 2 Fe atoms to approach within 20 Å. Where transfer is between two haem groups within the same protein (e.g. cyt b_{566} and cyt b_{562} in complex III, cyt a, and cyt a_3 in complex IV), these are fixed at a suitable distance from each other along the polypeptide chain (Section 12). Otherwise, **docking** mechanisms are necessary to align haem groups in different proteins.

In contrast, transfer of electrons from Fe to O_2 in haemoproteins

occurs with O_2 bound directly to the Fe atom, at a 'reserved' ligand binding site. An **inner sphere** transfer is said to occur. The mechanism of O_2 reduction reflects the chemistry of the dioxygen molecule itself, particularly (i) the thermodynamic requirement to carry out the initial reduction as a 2 e^- step, and (ii) the desirability of retaining highly reactive, partially reduced oxygen from the enzyme (see Section 9).

As an example, this section deals with our understanding of how these various considerations influence mechanisms of electron transfer in the cyt c_1-cyt c-cyt aa_3 system.

Electron transfer to and from cytochrome *c*
(Fig. 13.1)

As in most haem proteins, the haem moiety of cytochrome c is surrounded by protein, shielding its hydrophobic face from the solvent. Only one edge of the haem is exposed—and it is at this site that electrons must enter and leave the protein.

Electron transfer involves the following events.

1 *Orientation.* Mitochondrial cytochrome c diffuses in the intermembrane space, and along the outer surface of the inner membrane (Section 12). The protein has a very asymmetrical distribution of polar residues, leading to a high dipole moment with the positive end of the dipole at the haem edge. This will orient the protein so that the correct (haem) edge of the protein approaches the (negatively charged) cyt c_1.

2 *Docking.* Cytochrome c interacts with its redox partner cyt c_1, so that the haem groups of the two proteins are parallel. This is accomplished by a specific arrangement of lysine residues on cyt c (especially lys 13, 27 and 72) which flank the haem group and recognize a complementary pattern of acidic residues on the acceptor protein. Parallel haem rings allow interaction between the π electron clouds and ease electron transfer.

3 *Relaxation.* After docking, the two Fe atoms are about 17.5 Å apart. Computer simulations indicate, however, that the surface contacts are not rigid and the proteins can 'squash' together (within 1 ns) yielding a complex with the Fe atoms 15.5 Å apart. At the same time, phe 82, a conserved amino acid in cyt c, moves parallel to, and between, the haem rings. This possibly displaces solvent, promoting electron flow.

4 *Release and rotation.* Electron transfer to cyt c_1 then occurs, in the complex described above. For continued electron flow, cyt c must be re-oxidized, by its oxidizing partner, cyt aa_3. Acceptance of an electron from cyt c_1 involves the same face of cyt c (haem edge) and the same docking residues (lys) as electron removal by the oxidase. Thus the protein must be released from the cyt c_1, rotate and bind to cyt aa_3 for each catalytic turn over.

The acidic residues on cyt aa_3 that are involved in recognizing cyt c are on the Cu binding subunit II. In this case, the initial e^- acceptor on the oxidase appears to be Cu, not Fe.

The electrostatic nature of the recognition between cyt c and both its partners and the use of one constellation of groups to recognize these two (and other, unnatural, partners like cyt c peroxidase and cyt b_5) emphasizes the finding that specificity of recognition is fairly low. The recognition site is not sculptured precisely to match a single partner as in, for example, enzyme or antibody interactions. Such precise fitting may require rigidity in the proteins, whereas flexibility seems to be useful in the transfer mechanism (above).

Electron transfer to oxygen via cytochrome aa_3

Cytochrome aa_3 (complex IV) is a transmembrane protein with a large cytoplasmic domain where cyt c binds (Section 12). Figure 12.2 shows the arrangement, within complex IV, of the four metal cofactors. Haems a and a_3, and Cu_B are attached to the transmembrane helices of the largest subunit (subunit 1) of this complex; haems a_3 and Cu_B are buried some 10 Å inside the membrane while haem a lies close to the cytoplasmic surface. Subunit II, where it projects out of the membrane, binds the remaining copper (Cu_A), and also appears to be the recognition site for cyt c. The two haem groups are parallel to each other (for electron flow), and perpendicular to the membrane face. From this structural information, and from spectroscopic measurements of the cofactors, Fig. 13.2 gives the likely pathway for oxygen reduction.

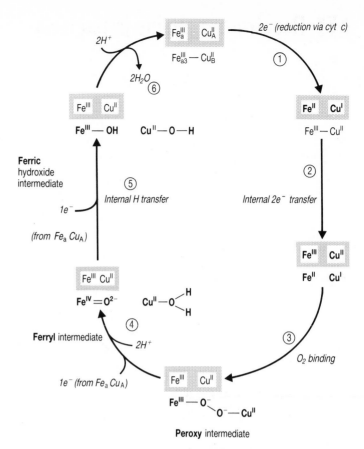

Fig. 13.2. Electron transfer in the cytochrome aa_3 complex (Complex IV).

1 Electrons enter from cytochrome c (see above) to haem a and Cu_A, at the cytoplasmic face. Both metals are reduced.
2 Two e^- are transferred to haem a_3 and Cu_B, revealing an O_2 binding site between the two. ($Fe^{III}a_3$ and Cu_B^{II} may be bonded together via a sulphur ligand; on reduction this bond is broken revealing a free ligand binding site for O_2.)
3 Oxygen binds and accepts $2e^-$ to form a peroxy intermediate. Transfer of one electron has been prevented—we have a $2e^-$ gate.
4 One more electron breaks the O–O bond, producing a tightly bound, ferryl (Fe^{IV}) oxide. The other oxygen is protonated and remains bound to Cu_B^{II}.
5 Further reduction results in ferric (Fe^{III}) and cupric (Cu^{II}) hydroxides.
6 The hydroxides are broken down by protons from the solution yielding H_2O and the original state of the a_3–Cu_B redox centre.

Note that, in accordance with our guidelines, O_2 is initially reduced by a $2e^-$ step, and the partially reduced forms of oxygen are always tightly bound to Fe of haem a_3. Also note that $1H^+$ is required per e^- transferred to oxygen.

14 Respiratory chains in bacteria

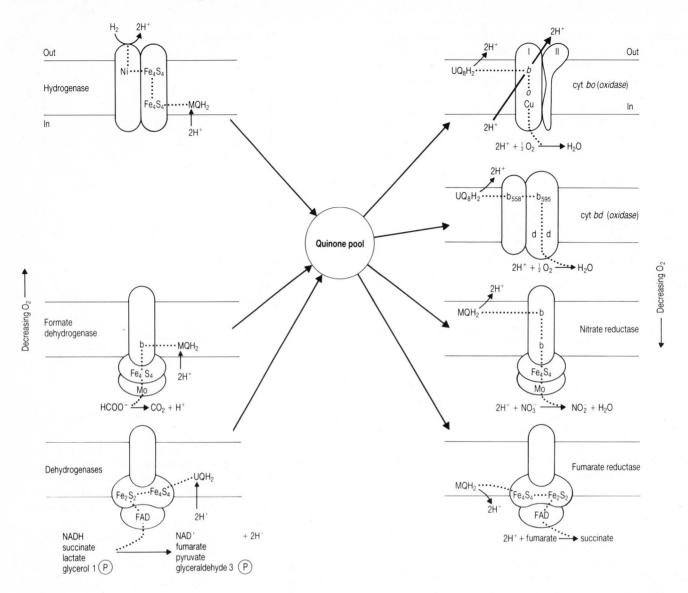

Fig. 14.1. The *E. coli* respiratory chain. Polypeptides bearing cofactors (and some anchor proteins) are shown. Dotted lines represent path of electron flow.

Dehydrogenases and oxidases

Oxidations *in vivo*, via a respiratory chain, can be considered to take place in two stages:

1 transfer of [**H**] from the reductant (in solution) to a quinone (in the lipid phase of a membrane); and

2 oxidation of the quinol by an **electron** acceptor.

Several dehydrogenases may be present, depending on the [H] donor; in mitochondria, for example, NADH, succinate and fatty acyl CoA dehydrogenases are present (Section 11). In bacteria, a *wider range of dehydrogenases* is found, reflecting the variety of available reductants (lactate, H_2, formate etc.).

Bacteria also employ a variety of quinol oxidases, as compared to the single pathway ($QH_2 \rightarrow$ cyt $bc_1 \rightarrow$ cyt $aa_3 \rightarrow O_2$) seen in mitochondria. This reflects not only the *variety of oxidants* that may be available (O_2, NO_3^-, SO_4^{2-} etc.), but also different strategies for the reduction of O_2 itself. Oxygen reduction invariably requires (i) a 5 liganded haem, where oxygen can bind and (ii) an initial 2 e^- step (i.e. the cooperation of two metal centres) (see Section 13). However, an *a*-type cytochrome (as seen in mitochondria) is not obligatory, and *various bacteria employ different classes of cytochrome* (*b, c, d*) in

O_2 *reduction*.

In addition, bacterial electron transfer chains are unlike mitochondrial chains in their **flexibility**. Different dehydrogenases or quinol oxidases can be induced or repressed, *depending on environmental conditions*, and slotted in as separate protein complexes (**modules**) communicating with the quinol pool. Mitochondria, in their constant environment, have a fixed chain composition and are unable to adapt in this way.

The *E. coli* electron transfer chain

A representation of the *E. coli* respiratory chain is given in Fig. 14.1 (not all the possible dehydrogenases are separately drawn). While following the pattern of (dehydrogenase complex)–(quinone pool)–(oxidase complex), the *E. coli* respiratory chain exemplifies a molecular organization very different from the mitochondrial one. Notably, *E. coli* contains no cytochromes *a* or *c*, and thus nothing analogous to mitochondrial complexes III and IV. In consequence, *quinol oxidation involves one, single transmembrane complex*, containing two *b* type haems and various other co-factors—depending on the final electron acceptor.

This diagram is tentative, partly due to metabolic variations in the chain making studies difficult, partly because a large fraction of *E. coli* cytochromes are of the *b* type and difficult to resolve spectroscopically, and partly because *E. coli* may also contain soluble dehydrogenases carrying out the same reactions (but without energy conservation). With our increasing knowledge of *E. coli* genetics, some of these problems should soon be resolved.

The reader should note that only a small number of these dehydrogenases, and only one of the quinol oxidases, will be active at any one time—depending on the environment. *The membrane will not contain all these components at once!*

Dehydrogenases

As in mitochondria, most *E. coli* dehydrogenases are *FeS flavoproteins* with their active sites on the internal surface of the membrane (inside the cell). The cofactors (FAD and several FeS centres) lie predominantly on this side of the membrane on two extrinsic membrane proteins non-covalently attached to an integral protein 'anchor'. The prevailing organization appears similar to that of mitochondrial succinate dehydrogenase (Complex II; Section 12), irrespective of the substrate involved.

A different type of organization is seen in formate dehydrogenase, where electrons from formate pass onto a molybendum cofactor, *molybdopterin*, and then, via an FeS centre, onto a cytochrome *b*. This transmembrane complex is similar in organization to nitrate reductase (see below).

In highly reducing environments, *E. coli* can obtain energy by oxidizing hydrogen gas (H_2). Its hydrogenase contains several FeS clusters and **nickel** (Ni, coordinated to cys residues) as a cofactor. Ni is a useful catalyst for reactions involving H_2 even in non-biological systems; here it presumably binds H_2 and oxidizes it in the first step of electron transfer.

Quinones

Two quinones may be involved in electron transfer in *E. coli*. One is a version of the mitochondrial *ubiquinone*, UQ_{10}, with a slightly shorter isoprene tail (UQ_8), and this occurs if oxygen is present. The other, **menaquinone** (a form of vitamin K), replaces UQ_8 in anaerobic electron transfer. Both dissolve in the lipid phase of the membrane, and accept/donate electrons to proteins in this phase.

Oxidases

E. coli may possess one of two possible oxidases—the cytochrome *o* complex (present when [O_2] is high), and the cytochrome *bd* complex (when [O_2] is low). Both contain electron carriers (haem *b* in cyt *o* complex, haem b_{558} and haem b_{595} in the cyt *bd* complex) which oxidize QH_2 near the outside (*periplasmic*) face of the membrane. Electrons then pass to a pair of oxygen binding cofactors (haem *o**, see Section 10, and Cu^{II} in the *o* complex, $2 \times$ haem *d* in the *bd* complex) near the internal face. This paired arrangement of carriers ensures $2\,e^-$ reduction of O_2 (Section 13).

The functions of cyt *o* and cyt *bd* complexes seem identical. However, cyt *bd* has the higher affinity for oxygen and *replaces cyt o in the membrane at low oxygen tension* (Fig. 14.1). There is also a difference in proton pumping ability; cyt *o*, like mitochondrial cyt *aa*, is a proton pump (Section 25), while cyt *bd* cannot pump H^+. This is related to the lower energy (ΔG) liberated by oxidations at very low [O_2] (Section 4).

Cyt *o* is thus analogous to mitochondrial cyt aa_3 (Complex IV). This is reflected in the amino acid sequence of its largest subunit, subunit I, which binds all three redox cofactors; this sequence is *homologous* to that of subunit I of cyt aa_3. Subunit II from cyt aa_3

* Haem *o* is closely related to haem *b*, except that one of the sidechains on the ring is modified.

and cyt *o* also show similarities in folding pattern, but sequence homology is small and, in particular, no Cu_A binding site is present in cyt *o*. This reflects the different source of electrons in the two complexes—cyt *o* is reduced by a quinone from the lipid phase, while cyt aa_3 receives electrons from (water soluble) cyt *c*.

Reductases

Under anaerobic conditions, *E. coli* may utilize respiratory chain electrons to reduce fumarate, an organic acceptor, or nitrate, an inorganic acceptor.

Perhaps not surprisingly, **fumarate reductase** and *succinate dehydrogenase* form a complementary pair of enzymes, either one or other being present. They are similar both in structure (sequence and cofactor content) and membrane organization (Fig. 14.1). However, their substrate *affinities* are widely different, the reductase having an affinity for fumarate some 20 times higher than the dehydrogenase.

Nitrate reductase comprises two extrinsic membrane proteins: α, containing *molybdopterin* and β, containing a Fe_4S_4 centre. These are associated with the transmembrane γ subunit, which bears two *b* type haems. The *b* cytochromes oxidize the quinol at the periplasmic membrane face and the electrons emerge onto NO_3^- via the molybdenum cofactor.

Interestingly, *nitrate reductase* is similar in organization to *formate dehydrogenase*, which operates in the opposite direction (to reduce quinone). A similar parallel was noted between fumarate reductase and succinate dehydrogenase (Fig. 14.1). The reason behind this structural 'pairing' is unknown.

Energy conservation

The two oxidases and nitrate reductase are reduced by quinol on the periplasmic side of the membrane, liberating H^+, and are subsequently oxidized on the cytoplasmic side, taking up H^+. A *proton gradient* is built up, and can serve as an energy store.

For many oxidations (e.g. lactate, glycerol-1-phosphate), this appears to be the only mechanism for creating an H^+ gradient in *E. coli* (i.e. the dehydrogenase complex is not energy conserving). It seems difficult for this organism to capture most of the energy released in the oxidation of, say, NADH by oxygen; efficient energy trapping seems to require a more complex respiratory chain, which includes (for example) two components, complexes III and IV, in quinol oxidation, and a more complex dehydrogenase structure. Energy trapping is discussed further in Section 19. In general terms, we might consider *E. coli* to have sacrificed '*efficiency*' of energy use for *flexibility* in the use of alternative electron donors and acceptors.

Other bacterial electron transfer chains

Variations between respiratory chains of different bacterial species may be of three types.

1 *Ability to use different electron donors and/or acceptors.* Dehydrogenases for organic compounds will have the same basic structure (FeS flavoprotein) as above, but others may require special cofactors (e.g. in the sulphur oxidizing bacteria H_2S is oxidized to S, and $S \rightarrow H_2SO_4$). Similarly, CO_2 or SO_4^{2-} may be used as a terminal electron acceptor in the methanogens or sulphur bacteria.

2 *Different organization within the chain.* Different species of bacteria may use (cyt *bc* + cyt aa_3), cyt *o*, or cyt *bd* to reduce O_2. Similarly, the structure of NADH dehydrogenase in some bacteria may be more complex (more like the mitochondrial dehydrogenase) than in *E. coli*. These changes affect the efficiency of energy conservation.

3 *The introduction of light-driven processes.* Photosynthetic bacteria (e.g. *Rhodobacter spheroides* or *Rhodospirillum rubrum*) can use an excited chlorophyll molecule to reduce the Q pool, and an oxidized chlorophyll to accept electrons from a bc complex. The nature of these photochemical reactions is discussed in Section 18.

15 Photosynthesis in green plants: 1 Path of electrons

Light and dark reactions

In green plant photosynthesis, a *poor oxidizing agent* ($NADP^+$) is reduced by a *very poor reducing agent* (H_2O) and, in a linked process, *ATP is made*. Both $NADP^+$ reduction and ATP synthesis require energy; this comes from the *absorption of light*.

These processes involve the internal (thylakoid) membranes of the chloroplast, and will be discussed in detail below. Further reactions

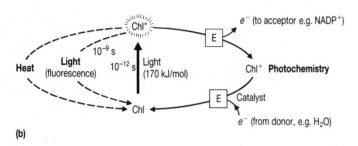

of NADPH and ATP (e.g. in the reduction of CO_2 to sugar phosphates or fatty acids) occur in the stroma of the chloroplasts. These constitute the 'dark reactions of photosynthesis' and are not dealt with here.

Chlorophyll and the trapping of light energy

Water, being colourless, cannot itself absorb light energy. Absorption is accomplished by the green dye, chlorophyll. Green plants contain two types of chlorophyll, *a* and *b*, both with a structure based on the chlorin ring (as in cytochrome *d*, Section 10). Mg^{2+} is the liganded metal (Fig. 15.1a).

In the light, chlorophyll absorbs light in the range 600–700 nm, and may then undergo a photochemical cycle (Fig. 15.1b, solid arrows). Chl*, the excited state, is a powerful *reducing* agent, capable of reducing, say, quinone. The oxidized form, Chl^+, which is thus generated, is a powerful *oxidant*. Chl^+ can remove [H] from water, and the result is a net transfer of electrons from H_2O to quinone. Overall, there is no net change in the chlorophyll molecule, which is restored to its Chl form. Chlorophyll is thus a *photosensitizer* in this process—it allows water to use the energy of a photon of light, but is not used up itself.

Chl* is very short lived in solution (lifetime $\approx 10^{-9}$s) and will readily re-emit the absorbed energy as fluorescence (Fig. 15.1b, dotted line). Furthermore, Chl^+ cannot easily remove electrons from water without the help of some catalyst (E). Thus for efficient photochemistry, chlorophyll must be integrated into an organized array, with nearby electron donor and acceptor—a **photosystem**. This is a *transmembrane complex* of several polypeptides, and is thus, structurally, analogous to the electron transfer complexes of mitochondria (Section 11).

In fact, a photosystem contains two classes of chlorophyll molecule. As its heart is the **reaction centre**, that chlorophyll which undergoes the photochemical cycle. Other molecules of chlorophyll absorb light energy, but pass it on to the reaction centre; they are called **antennae chlorophyll** and do not undergo photochemical changes themselves. Only the reaction centre Chl undergoes photochemical oxidation. Since the oxidized form, Chl^+, does not absorb red light, a *bleaching* (decrease in absorbance) is observed around 700 nm when the reaction centre is operative, due to the conversion of Chl to a steady-state level of Chl^+.

Two photosystems participate in green plant photosynthesis

Absorption of red light (around 700 nm) provides about 170 kJ of energy per mol photons. Thus the absorption of one photon can change the redox potential of chlorophyll by up to $-170/F = -1.4$ V. While, in principle, this is just sufficient to allow water ($E^{o'} \approx +0.8$ V) to reduce $NADP^+$, ($E^{o'} \approx -0.4$ V), it is insufficient in practice since (i) energy will be lost as heat, (ii) $NADP^+$ in the chloroplast will be more reduced than in its standard state, and (iii) some of the absorbed energy is needed for ATP synthesis. Thus, in the chloroplast membrane, *two photons cooperate in transferring each electron from water to NADP$^+$*

Cooperation between two photons requires the *sequential* participation of two photosystems—the first (**photosystem II**) removing electrons from water to produce a 'fairly good' reducing agent (plastoquinol), and the second (photosystem I) turning these into a 'very good' reducing agent (reduced FeS centres), which will reduce $NADP^+$. The reaction centres of the two photosystems are related in structure (see Section 16), but due to differences in environment they absorb maximally at different wavelengths. The reaction centre of photosystem II absorbs at 680 nm (**P680**), and that of photo-

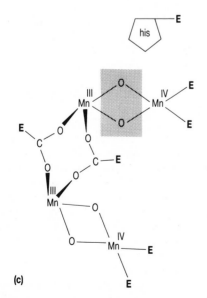

Fig. 15.1. (a) Structures of chlorophyll *a* and *b*.

(b) Photochemical cycle (solid arrows) and competing processes (dotted) in illuminated chlorophyll. E represents catalyst in photochemical processes.

(c) Possible structure of Mn cluster, showing oxygen atoms to be evolved as O_2 (shaded). E indicates remainder of protein, his a histidine residue. The oxidation state given = $2e^-$ removed in the $4e^-$ cycle.

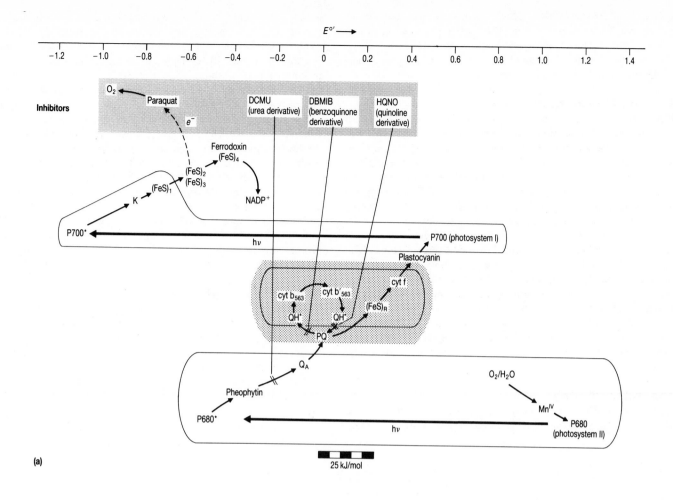

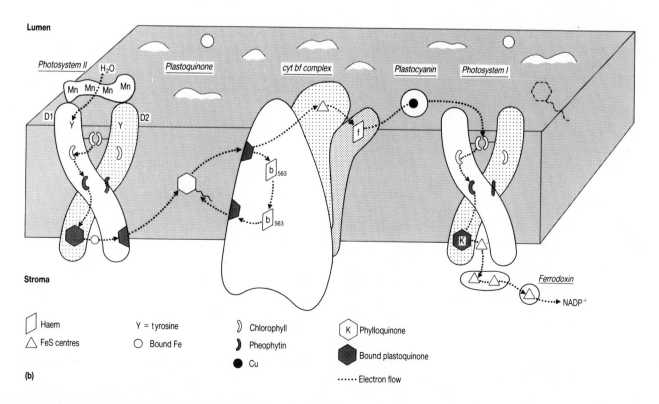

Fig. 15.2. **(a)** The electron transfer chain in chloroplasts. The enclosed cofactors are associated in membrane complexes (see Fig. 11.1). The central region (dotted) is similar to the central region of the mitochondrial chain.

(b) Membrane organization of the chloroplast electron transfer chain.

system I at 700 nm (**P700**). This accounts for the inability of chloroplasts to use light of 700 nm for efficient photosynthesis (the *red drop* phenomenon)—although photons at 700 nm can still be absorbed, they require the help of photons of slightly higher energy (lower wavelength) to oxidize water.

Electron flow in the chloroplast

Electrons are transferred from photosystem II to photosystem I by a series of membrane bound carriers—plastoquinone, a cyt *b/f* complex and plastocyanin. This pathway resembles the ubiquinone → cyt *c* segment of the mitochondrial respiratory chain (especially as cyt *f* is, in fact, a *c* type cytochrome) in composition (Fig. 15.2a) and membrane organization (Fig. 15.2b). This means that electrons traversing this path can pump protons (for energy conservation) while *increasing* their redox potential.

Hence, drawn on a scale of reducing power, electron flow in the chloroplast membrane is represented by the '**Z scheme**' (Fig. 15.2a, cf. Fig. 11.1). In this scheme, downhill electron flow occurs along the diagonal of the Z, with increasing redox potential (left to right on the diagram). Light driven (uphill) electron flow occurs along the horizontals (right to left). In this pathway, the transfer of electrons from PSI to NADPH occurs via FeS proteins, which are also familiar from the mitochondrial system. However, the **water splitting system**—which donates electrons to PSII—is unique to chloroplasts and is discussed below.

The water splitting system

Reduction of Chl$^+$, in photosystem II, requires electrons from water. The catalyst required (Fig. 15.1b) is (i) a cluster of 4 manganese atoms and (ii) a tyrosine residue (tyr$_z$), on a reaction centre peptide (see Section 16). The Mn atoms are *bound by carboxyl groups on the protein*, and also bind to *oxygen atoms from the water*, possibly as in Fig. 15.1c. They lie on the luminal surface of the membrane, close to the reaction centre chlorophyll (Fig. 15.2b).

Four electrons must be removed from two water molecules, and any partially oxidized intermediates tightly bound, before O_2 gas can be released. Conversely, one electron is needed to reduce Chl$^+$ coming from tyr$_z$. The manganese cluster thus provides a *gating mechanism*, and an oxygen binding site, to link these two processes (rather like cytochrome oxidase in O_2 reduction). This can be observed directly; if single electrons are removed sequentially from the reaction centre with brief flashes of light, O_2 is evolved only on every fourth flash.

Oxidation of water, by electron removal, will of course yield H^+ ions. The location of the water-splitting system means that these protons are liberated within the thylakoid lumen, contributing to the (coupling) H^+ gradient across this membrane (Section 24).

Paraquat as a herbicide

Photosystem I generates highly reducing (–ve $E^{0'}$) electrons, which are normally efficiently routed, through ferredoxin and NADP$^+$ reductase, to NADPH. Methyl viologen (paraquat) can also accept these electrons, and divert them to reduce oxygen (in a non-enzymic reaction) (Fig. 15.2a). Thus, highly reactive species such as $O_2^{-\cdot}$ and O_2^{2-} are produced in the chloroplast. These react with and damage, in particular, the highly unsaturated thylakoid lipids, killing the plant.

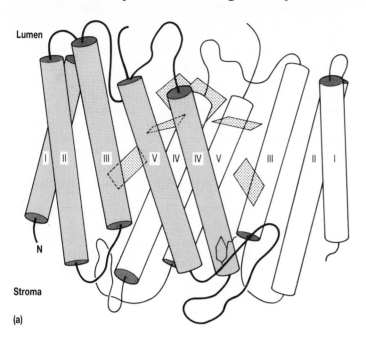

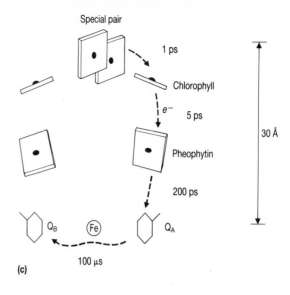

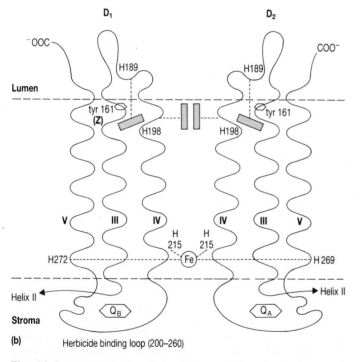

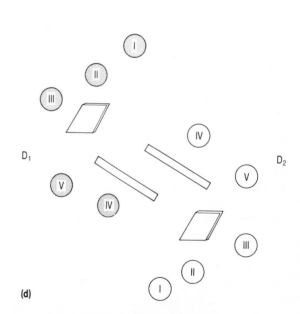

Fig. 16.1. **(a)** Organization of photosystem 2 (PSII) reaction centre in plants (based on structure of *Rps. viridis* reaction centre)—side view.

(b) Helices III–V of the reaction centre peptides D1, D2, showing functional residues. Q_A represents the tighter bound quinone moiety (pheophytin is not shown).

(c) Organization of the cofactors of the PSII reaction centre.

(d) PSII reaction centre viewed from luminal side of membrane.

Arrangement of proteins

Photosynthetic reaction centres comprise two similar (but not identical) polypeptides. In photosystem II of higher plants (D_1 and D_2), and in *Rhodobacter* spp. (L and M), the peptides are about 300 amino acids long. They are arranged in an (almost) symmetrical dimer, with each polypeptide containing *five* hydrophobic stretches, which cross the membrane as α helices. The N termini are on the stromal side in chloroplasts (Fig. 16.1a).

In these peptides, the *two C terminal helices* (IV, V) are specialized to bind the associated cofactors—**chlorophyll, pheophytin, quinones and iron**. At each end of helix IV, and at the stromal end of helix V are *histidine* residues, which coordinate magnesium (from the chlorophyll) and the associated iron atom. In the extramembrane loop between helices IV and V, there are conserved aromatic residues which interact with quinone (Fig. 16.1b).

In photosystem I of higher plants, the two transmembrane proteins (A_1 and A_2) are rather larger with additional transmembrane helices (700 amino acids). They still bind the cofactors involved in the photochemistry, arranged much as below. However, they also bind a total of about 40 molecules of antennae chlorophyll per pair of polypeptides.

Arrangement of the associated cofactors

In photosystem II, the cofactors are situated symmetrically in the two polypeptides. On the luminal side of the complex is a pair of closely apposed chlorophyll molecules, the **special pair**, nearly parallel to each other and perpendicular to the membrane plane. On the opposite side in a **single iron atom**. Both of these are bound by histidine ligands in both polypeptides and lie centrally in the dimer structure (Fig. 16.1b). These cofactors occur essentially inside a box constructed from the four helices (IV and V from each polypeptide) (Fig. 16.1d).

Within the membrane, but not perpendicular to it, are **two chlorophyll** and **two pheophytin** molecules, each liganded to one polypeptide chain but having a partner in an equivalent position on the other (Fig. 16.1c).

On the stromal side of the membrane, lying on the loop between helices IV and V and close to the iron atom, are **two quinone binding sites**—one on each chain. Thus, despite the non-identity of the two component polypeptides, the whole appears to comprise a nearly symmetrical arrangement. In fact, symmetry is broken as one quinone binding site (Q_A) appears **tighter** than the other; in the crystal only this quinone binding site, which is associated with the D2 polypeptide, is occupied.

A final, rather surprising, redox cofactor is a **tyrosine** residue of the polypeptide itself. This is designated electron donor Z (tyr_z) in the electron transfer pathway (Section 15). This amino acid is situated close to the special pair, on helix III in the membrane phase (tyr 161, on peptide D1). Again, there is a partner on the opposite polypeptide.

The *rigid framework* in which the cofactors are held, and the distances between them, are important for electron flow—allowing an electron to be removed from the special pair of chlorophyll molecules within 10^{-12} s, and to traverse the membrane within 10^{-9} s. This prevents loss of light energy by fluorescence (see Section 15).

Electron flow within the reaction centre

Light energy from antennae chlorophyll excites *only the special pair* of the reaction centre, due to its orientation perpendicular to the membrane (i.e. in the same direction as the antennae chlorophylls).

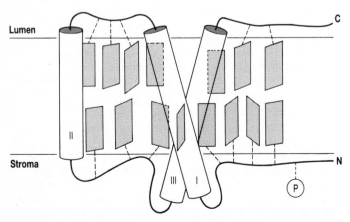

Fig. 16.2. Organization of antennae chlorophyll in plant light harvesting complex. Ⓟ represents the site of phosphorylation during regulation (Section 17).

Overlap of π electron clouds in the pair stabilizes Chl_2^+, easing the photochemical process.

Despite the almost symmetrical arrangement of the cofactors in the reaction centre, the electron passes across the membrane via only *one limb of the dimer*—on the side with the bound quinone (see Fig. 15.2b). On the way, the chlorophyll and pheophytin molecules are transiently reduced. The electron hole left on Chl_2^+ is filled by oxidizing the tyrosine residue (on the same side) which in turn oxidizes the Mn cluster (Section 15). The expelled electron arriving at the bound quinone, forms a semiquinone, perhaps sharing the electron with the iron atom. The electron remains here until a second electron arrives by the same route, after which a $2e^-$ reduction of the more loosely bound quinone (Q_B) occurs. Reduction of the second quinone takes much longer (some 10^{-4} s), and it marks the release of electrons from the complex into the lipid phase, and thus into the electron transfer chain.

Photosystem structure

This basic *reaction centre* structure, with its paired peptides and cofactors, is probably common to most photosynthetic systems. However, it must be adapted to accommodate the various electron donors and acceptors which vary between *photosystems*. This is accomplished by adding additional (generally peripheral) proteins (and cofactors) to the complex to mediate additional functions.

In the case of plant *photosystem II*, additional peptides are present on the luminal side of the membrane, a peripheral **33 kDa protein** at least being involved in stabilizing the manganese complex involved in water splitting (Fig. 15.2b). In the case of *photosystem 1*, the reaction centre peptides themselves bear a **4Fe4S centre** in place of the Fe liganded to histidine (above); in addition another FeS protein and an accessory protein are present on the stromal side, and a **19 kDa** plastocyanin binding protein is present on the luminal side (Fig. 15.2b).

In the *bacterial photosystem*, another transmembrane peptide (H) forms a cap over the quinone binding site, while on the periplasmic surface a **c-type cytochrome** is bound to a projecting arm of the M subunit.

The nature of the quinone also varies between photosystems. The bound quinone in PSII is **plastoquinone**; however, in photosystem I it is **phylloquinone** (vitamin K_1), and in the bacterial reaction centre, **menaquinone**. This may reflect, in part, the different redox potentials experienced by these electron acceptors in their respective complexes (see Fig. 15.2a).

Action of substituted urea herbicides

Substituted ureas, such as DCMU, prevent electrons leaving the reaction centre (Section 15). Studies with plants or bacterial strains resistant to these herbicides show the resistance mutations map in the IV–V loop of the (L or D_1) reaction centre peptide, close to the predicted 'loose' Q binding site (Q_B, Fig. 16.1b). It seems likely that these components are lethal due to their interference with Q_B binding.

Antennae chlorophyll

Only a small fraction of chlorophyll, in the membrane, lies in the reaction centre. Typically, *for each reaction centre, an additional 300–400 molecules of chlorophyll are present*. About 10% of these molecules are bound to the larger photosystem I peptides; the rest are bound in **light harvesting complexes**. Most of the light falling on a leaf is absorbed by these *antennae chlorophyll* molecules; the energy is rapidly (within 10^{-10} s) transferred to the reaction centre, which is the only site where photochemistry can occur.

Energy is transferred by **resonance energy transfer**, a short-range, dipole–dipole interaction between the chlorophyll rings. This requires alignment between the antennae chlorophyll molecules and the special pair at the reaction centre; thus the antennae chlorophylls must also be *perpendicular to the membrane face*. Note that, as the antennae chlorophylls are aligned with the special pair, they are not aligned with the other chlorophyll molecules of the reaction centre (Fig. 16.1c); these latter are involved in *electron* transfer, not *energy* transfer.

The light harvesting complex is a *trimer* of identical polypeptides, each containing *15 chlorophyll molecules*. Organization of each polypeptide is as indicated in Fig. 16.2, with three transmembrane helices, two of which protrude from the membrane on the stromal site. The chlorophyll molecules (7 on the luminal side, 8 on the stromal) 'hang' into the lipid phase from the region of peptide *along the membrane surfaces*, possibly through hydrogen bonds to the porphyrin side chains (and not via histidine ligation of Mg, as in the reaction centre, above).

The antennae chlorophyll serve two purposes. First, they *increase the light absorption by the membrane per electron transfer chain*. If, in normal daylight, one chlorophyll molecule absorbed a photon on average each second, while electron transfer took 20 ms to complete, a 1 : 1 Chl : electron transfer chain ratio would mean that each chain was idle 99.8% of the time. It is thus more efficient to pool absorbed energy from about 500 chlorophylls (in this example) and synthesize fewer respiratory chains. This is borne out by observations on plants adapted to strong light (which have fewer antennae chlorophyll) or shade (which have more).

Second, the antennae *broaden the wavelength range* over which the leaf can absorb light. There are two types of chlorophyll (*a* and *b*) in the antennae and different molecules lie in different environments within the proteins, inducing a variety of absorption maxima over the range 600–700 nm. Thus, rather than showing maximal photosynthetic efficiency around 680–700 nm (where PSI and PSII absorb), photosynthesis is equally efficient over a wide range of wavelengths.

17 Photosynthesis in green plants: 3 Organization of membranes

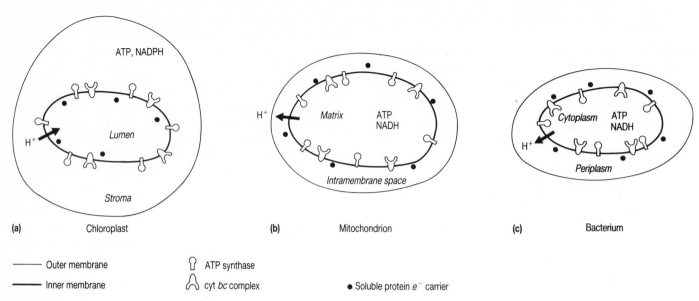

Outer membrane —— ⏝ ATP synthase

Inner membrane —— ⏝ cyt *bc* complex

● Soluble protein *e*⁻ carrier

Fig. 17.1. Orientation of coupling membranes in different systems.

Transmembrane asymmetry

Biological membranes typically show **transverse asymmetry**; one face has a different structure (and hence function) from the other. This is true of the chloroplast membrane. Notably (i) the water splitting system and the small mobile protein, plastocyanin, are on the luminal side of the membrane; (ii) the $NADP^+$ reductase is on the stromal side of the membrane; and (iii) the electron transfer complexes (PSI, PSII, and cyt *bf*) have suitable cofactor orientations (see Fig. 15.2). In addition, the ATP synthase (Section 26) has its catalytic domain on the stromal side of the membrane—which is where ATP is made. This organization requires proton translocation, driven by electron transfer, to deposit H^+ *inside the lumen* of the thylakoids, whence they move outwards to drive ATP synthesis (Fig. 17.1a).

This orientation contrasts with that of mitochondria and bacteria which, despite having analogous transmembrane complexes, show the opposite orientation. In these cases, the small mobile electron carrier protein, cyt *c* in mitochondria, is located outside the closed vesicle, with NADH oxidation and ATP synthesis occurring inside. H^+ ions must be pumped *outwards* to be available for ATP synthesis. These contrasting orientations are displayed in Fig. 17.1.

When considered *relative to the orientation of the H^+ gradient*, however, *the orientation of proteins within coupling membranes is constant*. This is due to the common role of the H^+ gradient in ATP synthesis (Section 20). For example, cyt c_1 and cyt *f* both protrude from the membrane face at low pH (high H^+), while ATP is made in the compartment at higher pH (Fig. 17.1). For convenience, the high H^+ side is termed the P side (Positive) and the low H^+ side, the N side (Negative). *Membranes and membrane proteins in this text are all depicted P side uppermost*, for ease of comparison.

Lipid composition

Like all biological membranes, chloroplast thylakoid membranes comprise a number of transmembrane protein complexes embedded in a lipid bilayer. Unlike most membranes, however, the thylakoid lipids are largely (> 75%) uncharged **galactolipids**.

In galactolipids (Fig. 17.2), the *glycerol* moiety is joined to two *fatty acids* by ester links (as usual), but the third – OH group is linked via a glycosidic bond to the sugar, *galactose*. This yields a lipid with a polar, *but unchanged*, headgroup—unlike the more common *phospho*lipids). The predominance of monogalactosyl diacyl glycerol (Fig. 17.2) and its digalactosyl analogue leads to a *very low charge density at the surface* of chloroplast membranes. This is important for the close stacking of thylakoid membranes (see below).

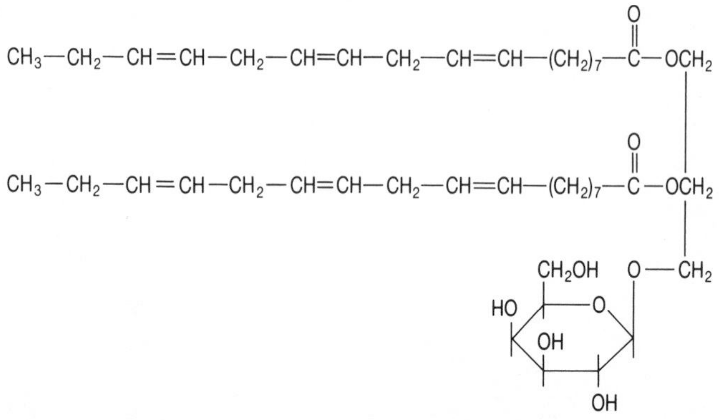

Fig. 17.2. Monogalactosyl dilinolenoyl glycerol.

Another feature of chloroplast lipids is their *high degree of unsaturation*; typically 3 $C=C$ double bonds are present (Fig. 17.2). Thus the membranes are highly fluid, even at low temperatures (freezing point = –30°C). This is essential for function—plastoquinone must move rapidly in the membrane phase (see below), and plants of course do not maintain a high 'body temperature' like the higher animals.

Lateral asymmetry

As is clear from electron microscopy, thylakoid membranes are closely apposed in some regions (**grana stacks**) and not in others (**lamellae**). This structural specialization is reflected in differences in composition; the stacked regions contain most of the PSII and the light harvesting complex, and the lamellae all the PSI and ATP

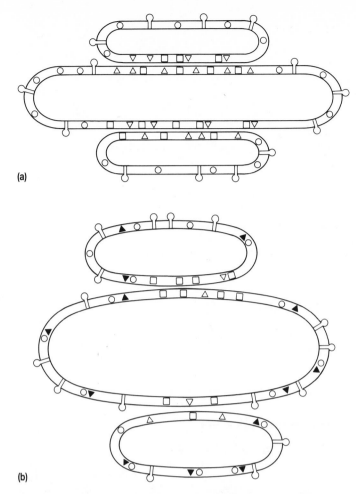

(a)

(b)

Fig. 17.3. **(a)** Distribution of PSI (◯), PSII (▢), light harvesting complex (△) and ATP synthase ⨄ in stacked and unstacked regions of thylakoids.

(b) Change in distribution on illumination with light of < 680 nm (note also some degree of unstacking). Symbols as (a); (▲ = LHC-ⓅP). (Cyt *bf* complex is evenly distributed, and not shown here.)

synthase (Fig. 17.3a). (The cytochrome *bf* complex appears randomly distributed.) Thus the membrane shows **lateral asymmetry** (differences in composition from place to place) as well as *transverse asymmetry* (differences in composition on opposite sides).

Clearly, for electrons to be transferred from PSII to PSI, in the Z scheme, one or more electron carriers must be able to shuttle electrons between the grana stacks and lamellae rapidly enough to account for the observed rates of photosynthesis (about once every 10 ms). The large transmembrane complexes, and probably even plastocyanin, cannot move fast enough to allow such a collision rate (see below); however, *plastoquinone is highly mobile within the membrane phase, and appears to be the electron shuttle*. Plastoquinone can move 100–1000 nm in 10 ms, while the mean distance between granal and lamellal complexes is 20–200 nm.

Distribution of energy between photosystems— the carburettor effect

For the Z scheme to operate, photosystems I and II must operate at identical rates. However, since they have different adsorption maxima (700 nm, 680 nm respectively) at any randomly chosen wavelength, we would expect one to work faster than the other (for example at 650 nm, PSII > PSI).

The chloroplast overcomes this problem by adjusting the distribution of antennae chlorophyll; if photosystem II is working too fast, antennae chlorophyll are moved from PSII to form an association with photosystem I and thus increase its absorption of photons. To accomplish this, the trimeric *light harvesting complexes (LHC) (Section 16) move away* from PSII (in the grana) and close to PSI (in the lamellae) (Fig. 17.3b). Diffusion of these large protein complexes through the membrane (despite its highly fluid lipid) is slow, and *this adaptation takes several minutes*; this is presumably adequate in the slowly changing environment of the plant.

Movement of LHC is triggered by the *redox state of plastoquinone*—if PSII works faster than PSI, quinone will become highly reduced because electrons cannot be removed fast enough (see Fig. 15.2a). This change is sensed by a **protein kinase** which phosphorylates the polypeptides of the light harvesting complex (producing LHC-ⓅP). Phosphorylation occurs on the stromal side of the membrane, close to the N terminal end of the polypeptide (see Fig. 16.2). Phosphorylation lowers the affinity of LHC for PSII, and increases it for PSI; in addition, the introduction of *negative charge* into the highly apposed grana membranes leads to repulsion (see above) and an outward movement of LHC-ⓅP towards the lamellae and PSI. Thus PSI *acquires more antennae chlorophyll* and the balance of energy trapping is restored.

Phosphate is removed from LHC-ⓅP by a **phosphatase**; if the quality of light changes and PSI is now outstripping PSII, the level of reduced quinone falls, the kinase becomes less active and the phosphatase reduces the steady state level of LHC-ⓅP, thus redistributing more of the antennae chlorophyll back towards PSII.

This model rationalizes the existence of lateral asymmetry in the chloroplast membrane, and also its unusual content of uncharged lipids. Presumably, for charge repulsion to drive redistribution, an uncharged membrane surface is necessary.

Evidence for wavelength compensation by redistribution of antennae

Chlorophyll fluorescence is an indicator of inefficiency *in vivo*

As noted in Section 15, the reaction centre is adapted to *trap light energy in photochemistry*, rather than *lose it by fluorescence* emission. If photochemistry is blocked, however, the absorbed energy is re-emitted by the antennae chlorophyll as fluorescence. This occurs if PSII is working faster than PSI—quinone is reduced and an excited special pair cannot lose its electrons.

Thus we have a useful *probe* for the redistribution of LHC during adaptation to light of a given wavelength. Excitation at 650 nm, as described above, initially produces high fluorescence because PSII is overexcited relative to PSI; however, as LHC moves to equalize excitation of PSI (Fig. 17.3b), fluorescence falls to a minimal value (indicating maximum efficiency of photosynthesis).

18 Photosynthesis in bacteria

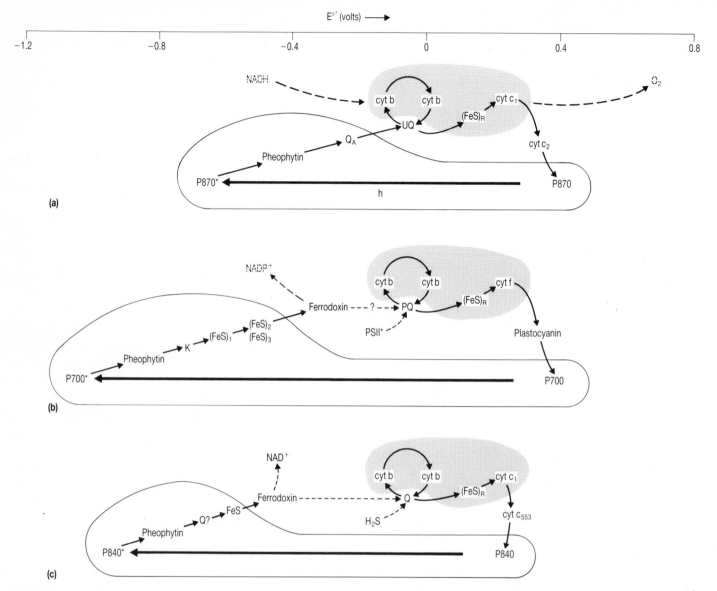

Fig. 18.1. Cyclic electron flow in photosynthesis. In each case, a source/sink for electrons—as is employed in linear electron flow—is shown in broken type. (a) Rhodobacter, (b) plant photosystem I, (c) chlorobium.

Principles of photosynthesis

The essential features of plant photosynthesis are:

1 *light driven oxidation of chlorophyll*, generating electrons at a highly negative redox potential (highly reducing) and oxidized chlorophyll;

2 *reduction of oxidized chlorophyll* by electrons at a positive redox potential;

3 *generation*, during these processes, *of a transmembrane H^+ gradient* (allowing the absorbed energy to be trapped as ATP).

These features are maintained in bacterial photosynthesis. However, bacteria are adapted to utilize a wider variety of environments than plants.

1 Some species (e.g. the 'purple' bacteria) will utilize *different wavelengths* of light to green plants (> 750 nm), and thus can live in muddy ponds, etc.

2 Some species utilize *reductants other than water*—H_2S, organic compounds or even, in cyclic systems, the electron released from the reaction centre itself.

In addition, bacteria have no specialized chloroplasts for photosynthesis; *photosystems and electron transfer complexes are present in the plasma membrane* (cf. Section 14). This requires some modification to the arrangement of the electron transfer chain and/or the antennae pigments of the photosystems.

Patterns of electron flow

Cyclic electron flow

Possibly the simplest photosynthetic arrangement is seen in the **purple (non-sulphur) bacteria** such as *Rhodospirillum, Rhodobacter* (Fig. 18.1a). The membrane contains a reaction centre whose structure (and action) is very similar to chloroplasts PSII (Section 16). Indeed, our model of the plant system derives from structural studies on a purple bacterial reaction centre. The *special pair* of bacteriochlorophyll molecules is excited by light of considerably longer wavelength than in green plants (**P870**), but nonetheless the electron crosses the membrane, emerging on *ubiquinone* (not plastoquinone) at the cytoplasmic membrane face.

The quinol is then oxidized by a transmembrane cyt bc_1 complex, and the electron re-emerges on the periplasmic side to reduce an iron–sulphur protein, cyt c_2 and ultimately $P870^+$, the oxidized reaction centre chlorophyll. *No net reductant is generated* (the electron released from P870 is used to reduce $P870^+$), but a H^+ gradient is generated across the membrane. The membrane organization of these complexes is shown in Fig. 18.2

The analogy between this electron transfer system and the $UQ \rightarrow$ cyt c span in mitochondria is obvious (see Section 12)—it is as if a 'photosystem module' and a 'cyt bc_1 complex' module have been inserted into the same membrane with a suitable adaptor (cyt c_2) at the membrane surface to allow intercomplex transfer. This idea (see Section 14) is emphasized in these particular bacteria,

which are *facultative aerobes*; if they are grown aerobically, the photosystem disappears but the cyt bc_1 complex is recruited into a (mitochondrial-like) respiratory chain which oxidizes NADH (Fig. 18.1a).

Cyclic electron flow also occurs in some green bacteria (e.g. *Chromatium*), although the alternative, aerobic mode of life is absent in these species. It also occurs in chloroplasts around PSI; electrons from P700*, passing through ferrodoxin can reduce the cyt bf complex and thence via plastocyanin, reduce $P700^+$ (Fig. 18.1b). (How the electrons are routed from ferrodoxin to cyt bf is unknown.) Chloroplasts can thus, if necessary, *increase ATP production* (cyclic and non-cyclic electron flow) *relative to NADPH* (non-cyclic flow only).

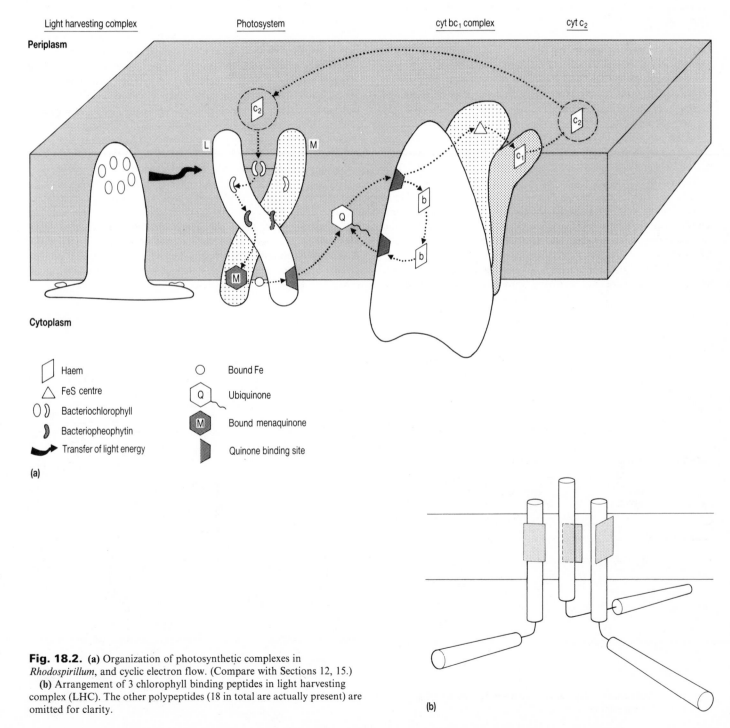

Fig. 18.2. **(a)** Organization of photosynthetic complexes in *Rhodospirillum*, and cyclic electron flow. (Compare with Sections 12, 15.)

(b) Arrangement of 3 chlorophyll binding peptides in light harvesting complex (LHC). The other polypeptides (18 in total are actually present) are omitted for clarity.

Linear electron flow with one photosystem

Like plants, some bacteria (the **green, sulphur bacteria** such as *Chlorobium*) can use electrons from an excited photosystem to reduce $NAD(P)^{+\dagger}$. However these bacteria utilize only one photosystem; as a result they are unable to oxidize water, requiring a stronger reductant such as H_2S (yielding sulphur, Fig. 18.1c) or even an organic substrate. They can also carry out cyclic electron flow (see above) for ATP synthesis.

Linear electron flow with two photosystems

The cyanobacteria (e.g. *Anabaena*, *Synechococcus*), besides being capable of the two above modes of energy generation, can also carry out oxygenic photosynthesis like green plants. This requires the cooperation of the two photosystems and, again like green plants, these absorb light at around 680 nm and 700 nm. (Again like plants, but unlike bacteria, they utilize *plastoquinone*—not ubiquinone or menaquinone—as the mobile electron carrier.) This arrangement is shown in Fig. 15.2.

Light harvesting systems

Photosynthetic bacteria, like chloroplasts, contain accessory pigments which absorb light and pass their energy on to the reaction centre (see Sections 16, 17). However, although the reaction centres in plants and bacteria are similar, the organization of these *antennae* pigments is very different.

In the purple bacteria, the light harvesting systems (LHC) are complexes of 18 short (approximately 50 amino acid) polypeptides, each of which binds one molecule of bacteriochlorophyll (BChl) on its *single transmembrane helix*. Mg^{2+} within the membrane, *at the same level as the 'special pair' of the reaction centre* (Fig. 18.2), which allows energy transfer between them. This is rather a different arrangement to the plant LHC, where chlorophyll was bound to the surface regions of the peptide, and not to the transmembrane helices (Section 16).

Bacteria may also contain light harvesting assemblies *on their membrane surfaces* i.e. not involving transmembrane proteins. The green bacteria contain chlorosomes, containing many thousand BChl molecules, on the cytoplasmic membrane face. The cyanobacteria contain phycobilosomes, which contain large numbers of *phycobilins* (linear tetrapyrroles) bound to peripheral membrane proteins in a semi-crystalline array.

Evolution of photosynthetic systems

Although plant PSI and PSII are similar (special pair, quinone acceptor etc.), there are considerable differences between them. Structurally, *the 2 reaction centre polypeptides of PSI are much larger* than those of PSII (binding additional Chl molecules), and in PSI, electrons leave the complex via an *iron–sulphur centre* (rather than a quinone in PSII). Functionally, PSI operates over a *more negative redox span* than PSII (Section 15).

We have already noted strong similarities between *plant PSII and the reaction centre of purple bacteria*. From the comparative redox spans (Fig. 18.1) and the presence of FeS centres in the reaction centres of the green bacteria, it seems likely that a homology exists between *plant PSI and the green bacteria reaction centres*. This has been confirmed by sequencing the reaction centre proteins.

Oxygenic photosynthesis (that releasing O_2) presumably evolved through a series of steps of increasing complexity. Incorporation of his process into a eukaryotic cell occurred, almost certainly, when a primordial eukaryote engulfed a free living photosynthetic bacterium (the 'endosymbiont' theory; see Section 45). The cyanobacterial membrane carries out oxygenic photosynthesis and is similar in organization to (was probably the ancestor of) the red algal *chloroplast* in possessing chlorophyll *a* and phycobilins (see above).

Higher plant chloroplasts, in contrast, contain *both chlorophyll a and b*, organized in light harvesting, *transmembrane* complexes (Section 16). They may have evolved, not from the classical cyanobacteria, but from the **prochlorophytes** (*Prochloron*, *Prochlorothrix*), photosynthetic bacteria with both chlorophyll *a* and *b* (and no phycobilins). Sequencing shows that photosystem peptide D1 in *Prochlorothrix* resembles that of higher plant chloroplasts, more than that of the cyanobacteria.

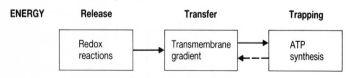

Fig. 19.1. 'Central dogma' of energy conservation.

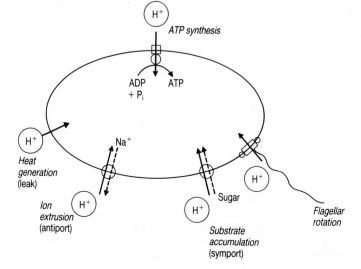

Fig. 19.2. Possible roles of H^+ gradient as an energy source (solid arrows show downhill movement of H^+).

19 Energy transduction

Redox energy to anhydride bond formation

The production of ATP requires energy to be transferred *from one type of chemical reaction (oxidation) to another (anhydride bond formation)*. In substrate level phosphorylation (see Section 8), the transducer is typically an enzyme in which oxidation is used to create a thioester bond. Thiol (-SH) groups make the link because they can participate in both redox reactions and anhydride (thioester) formation.

No such transducer can participate in respiration-linked phosphorylation since, within the membrane, redox reactions and ATP synthesis involve separate enzyme complexes. No single complex links the two reaction types. This means there must be a separate **energy transfer** step, interposed between the *energy releasing* oxidations and the *energy utilizing* dehydration—see Fig. 19.1, facing page.

(Scheme 19.1 incidentally mirrors the outline of the central section of this book; Sections 11–18 deal with energy release, Sections 19–25 with energy transfer mechanisms and Sections 26–30 with ATP synthesis.)

The chemiosmotic theory

It was Peter Mitchell (Nobel laureate for chemistry, 1978) who, in the early 1960s, realized the nature of the energy transfer mechanism. He suggested that

1 The energy transfer intermediate, linking oxidations and ATP synthesis is a *transmembrane ion gradient*.

2 The ions involved in the gradient are *protons* (H$^+$).

These form the essential postulates of the chemiosmotic theory. The name 'chemiosmosis' signifies the link between *chemical* reactions and energy stored in a transmembrane gradient ('*osmotic* energy').

The logic of chemiosmosis

We have seen (Section 5) that energy can, indeed, be stored in an ion gradient—*providing that the ions are not at equilibrium*. The energy stored in a gradient of univalent cations is given by equation 7.1

$$\Delta \overline{G}_{A \to B} = \Delta \tilde{\mu}_{x^+} = RT \ln \frac{[X^+]_B}{[X^+]_A} + F(\psi_B - \psi_A) \; (J/mol \; X^+)$$

i.e. the difference in (electro)chemical potential of the ions on opposite sides of the membrane. Note that the energy term has two components—a concentration gradient and a charge gradient—which reinforce each other (see Section 7). There appears to be no restriction as to which term might predominate *in vivo*—in mitochondria the electrical component is much larger than the concentration term, while in chloroplasts the opposite is true.

This equation emphasizes the fact that *energy stored in a gradient is independent of the chemical nature of the ion participating*. Hence the second postulate, which recognizes H$^+$ as the particular ion involved. H$^+$ is a convenient ion; it can be readily transmitted through water, it has well defined binding sites on proteins (acid–base groups) and, perhaps most importantly, it is a participant in many redox reactions (Section 10).

Contrary to initial expectations, the *chemical nature* of H$^+$ as an energy transfer intermediate is not directly relevant to the *chemistry* of the ATP synthesis reaction. ATPases which pump a wide variety of ions (see Section 43) can be reversed to make ATP. The ATP synthase of chloroplasts and mitochondria is adapted to use protons as a store of energy simply 'because they were there'.

Consequences of the chemiosmotic postulates

These postulates mean that three requirements must be met by the respiratory chain.

1 *Redox reactions within the chain must transfer H$^+$ across a membrane.* Chemically, this is not a simple problem. A cyclic chemical (**scalar**) process (oxidized → reduced → oxidized form)—leading to no net change in a respiratory carrier—must nonetheless carry out a **vectorial process** leading to a net movement of protons from one place to another.

2 *The enzyme, ATP synthase, must use the transmembrane gradient of protons to make ATP.* Again note that ATP synthesis occurs on one side of the membrane only—it is a scalar process—while the ions are moved *across the membrane*.

3 *The redox reactions and ATP hydrolysis (reversal of synthesis) must require H$^+$ translocation*. In the terms of Section 8, chemical reaction and ion pumping must be **coupled** together. If either chemical reaction could occur without ion pumping, the energy released could not be used and only heat would result.

Note that the chemiosmotic postulates do not make any stipulation of the *mechanism* of ion pumping or gradient utilization—they simply require such mechanisms to exist.

Delocalized protons

It is implied by the above postulates that the protons moved by the respiratory chain are *free to diffuse in the aqueous phases* on either side of the membrane (or, if bound to proteins, are in equilibrium with these phases). The protons are said to be **delocalized**—they can move from one part of the membrane (e.g. a respiratory complex) to another (an ATP synthase) rapidly, at diffusion limited rates. Only in this situation will the above calculation of stored energy hold.

This model requires the membrane to be a *passive barrier to H$^+$ diffusion*—and indeed protons do not diffuse easily across the lipid bilayer of membranes. It also means that coupling between oxidation and ATP synthesis is *indirect*—the H$^+$ gradient may be 'hijacked' by other transmembrane proteins and its energy used to drive other processes. In the mitochondrion, these mainly comprise *ion transport systems* (P$_1$ transport is linked to H$^+$ movement, for example, Section 31); in bacteria, flagellar movement (and thus movement of the whole cell) would be included (Sections 38–39). A generalized model, indicating the possible uses of delocalized protons, is given in Fig. 19.2, facing page.

The implications that coupling protons are delocalized—and in particular that *all* such protons are *always* delocalized—has been one of the most contentious aspects of the chemiosmotic theory, and is dealt with in more detail in Section 32. For the present, we shall assume that, at least usually, the protons involved in energy transfer are indeed free in solution.

20 Energy transfer via a H⁺ gradient—testing the model

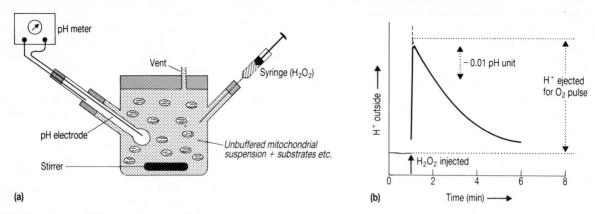

Fig. 20.1. Measurement of H⁺ extrusion in mitochondria given an oxygen pulse.
(a) Apparatus.
(b) Results. Upward limb—extrusion of H⁺ during oxidation. Downward limb—leakage of H⁺ back into mitochondria after exhausting oxygen.

Does the respiratory chain pump protons?

The first requirement for the operation of chemiosmosis is that electron transfer chains must build up a transmembrane gradient of protons (Section 19). The ability of the mitochondrial redox chain to act as a H⁺ pump was classically tested in the **oxygen pulse experiment** (Fig. 20.1a). Mitochondria are incubated with a substrate (e.g. β-hydroxybutyrate, which produces NADH inside the mitochondrion) and allowed to consume all the available oxygen. The pH of the solution (outside the mitochondrion) is monitored with a glass electrode. This gives the initial (baseline) pH of the solution.

A pulse of H_2O_2 (which will provide O_2) is then injected into the mitochondrial suspension. As can be seen in Fig. 20.1b, the pulse of oxygen leads to a rapid appearance of H⁺ outside the mitochondrion, which is slowly reversed after all the oxygen is consumed. This is interpreted as *redox reactions pumping protons from inside the mitochondria to the outside*; when the redox reactions cease, the protons slowly leak downhill, back across the membrane. Thus the redox chain *can* pump protons.

Quantitation

From this type of experiment, estimates can be made of the **H⁺/O ratio**—the number of protons transferred per O atom reduced. For accurate measurements, care must be taken that *a known amount of added oxygen* (conveniently all O_2 added) is consumed, and that *all protons* pumped are seen by the electrode. Precautions necessary to achieve these ends are explained in Section 23.

Under these conditions, H⁺/O ratios can be calculated from the *height of the peak*, corrected for leakage, divided by total O added. Values of 10–12 H⁺ released per O consumed, using β-hydroxybutyrate as substrate, have been recorded. A less reducing substrate, like succinate, gives H⁺/O ratios of 6–8.

Can the ATP synthase use a proton gradient?

The second requirement for the operation of chemiosmosis is that the ATP synthase must be able to use the proton gradient for ATP synthesis. The approach here has been to impose, artificially (i.e. without electron flow), a pH gradient across mitochondrial or chloroplast membranes—and then to see if these 'energized' membranes can synthesize ATP from ADP + Pᵢ.

The classical experiments were performed on chloroplast thylakoid membranes. Electron flow is inhibited, and the vesicles soaked in buffer at pH 5.0 ('**acid bath**'). After soaking for several seconds

the internal pH (in the lumen) is low. The vesicles are then transferred to a buffer at pH 8.5, containing ADP + [³²P]Pᵢ. *ATP is formed as the internal protons move outwards towards the high external pH*. Thus the ATP synthase *can* use energy from a H⁺ gradient.

A similar experiment, using mitochondrial membrane vesicles, is depicted in Fig. 20.2.

Supporting experiments

The above experiments **directly test** the requirements of the chemiosmotic theory. Besides performing these tests on a variety of systems (mitochondria, chloroplasts and bacterial membranes), a number of other experiments **support its predictions**:

1 An intact membrane is necessary to provide a barrier to proton leakage.
Test Respiration-linked ATP synthesis is inhibited by very low concentrations of detergent, which make membranes leaky without (apparently) damaging proteins.
2 No *specific* contact is needed between energy generator (the redox complex) and energy user (the ATP synthase).
Test Incorporation of an isolated ATP synthase from one species into phospholipid vesicles with a proton pump from a different species leads to a system capable of ATP synthesis. A variety of proton pumps (cytochrome oxidase, bacteriorhodopsin etc.) have been used in such experiments.
3 Since protons are pumped into an aqueous space, there will be no fixed relationship between the number of electron carrier molecules (or ATP synthase) and the amount of energy stored.

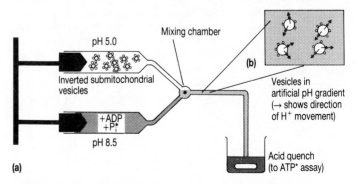

Fig. 20.2. Rate of ATP synthesis after creation of a pH gradient.
(a) Quench flow apparatus; ATP is made during flow from mixing chamber to quenching solution (5–100 ms).
(b) pH gradient set-up after mixing.

Test Chloroplasts preilluminated without ADP + P$_i$ can make ATP if these are provided in a following dark period. The amount of ATP made is 50–100 times higher (mol/mol) than any electron carrier. (The chloroplasts presumably accumulate H$^+$ in the illumination period.)

This experiment also shows that ATP synthesis and oxidations can be separated from each other in time—the connection between them is indirect.

Kinetic competence

It is not sufficient just to show that redox reactions *are capable of* building up a proton gradient, or that protons *can* drive ATP synthesis. It must also be established that they do so **fast enough** to account for rates seen *in vivo* (and hence are not just on a minor side pathway of a true coupling mechanism). To do this, we require techniques for measuring processes such as ATP synthesis or H$^+$ fluxes very rapidly—over periods of 1 s or less. Special rapid mixing systems, single turnover flashlamps (in chloroplasts), and spectral probes for pH (not electrodes, which respond too slowly), must be used.

Can ATP be made fast enough?

Rates of ATP synthesis can be measured by modification of the acid bath experiment (above) to include a quench flow system. In one variation (Fig. 20.2) inverted submitochondrial vesicles loaded internally with protons at pH 5.0 (in one syringe), are forced into a mixing chamber with a strong buffer at pH 8.5. Mixing takes less than 5 ms. The second buffer contains ADP + [^{32}P]P$_i$, and as the reagents pass down the tube, ATP is made. At the end of the tube, the reagents flow into cold acid which stops the reaction. By varying the length of the reaction tube, the time allowed for reaction can be varied between 20 ms and a few seconds.

If this system is optimized (by blocking electron transfer and including K$^+$/valinomycin to prevent charge imbalance, see Section 22), rates of phosphorylation are as fast as when phosphorylation is driven in the conventional way using NADH as substrate. Thus (delocalized) protons can indeed drive phosphorylation at normal rates—*a proton gradient is kinetically competent to act as an energy transfer intermediate*. Similar experiments have been performed on chloroplasts.

Can protons be pumped fast enough?

To measure rates of proton transfer, we need (i) to start electron flow very rapidly, and (ii) to be able to measure protons, both inside and outside membrane vesicles, over very short periods. The first requirement limits us to photosynthetic systems, where electron transfer can be initiated by brief flashes of light. The second requires the use of *pH indicators*, which respond rapidly to pH changes and can be monitored spectrophotometrically. (pH electrodes have response times in the range of 0.1–10s, and, in addition, are unable to probe inside vesicles.) The experimental set-up is thus very different from that in O$_2$ pulse experiments.

To measure external pH in a suspension of chloroplasts (Fig. 20.3a), the charged (impermeant) indicator *cresol red* was used. After a 10 μs light flash (which turns over the electron transfer chain once), the extravesicular pH rises (i.e. protons disappear outside the

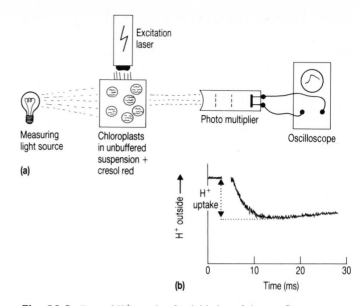

Fig. 20.3. Rate of H$^+$ uptake after initiation of electron flow.
(a) Chloroplasts excited by laser flash; H$^+$ uptake followed by absorption of cresol red indicator.
(b) Time course of H$^+$ uptake by chloroplasts.

vesicles) within about 1 ms (Fig. 20.3b). Since the chloroplast ATP synthase turns over about once every 20 ms, *proton pumping is sufficiently fast to lie on the energy transfer pathway to ATP synthesis*.

To measure internal pH the *permeant indicator*, neutral red, was used, together with an *impermeant buffer* (albumin). The neutral red outside the vesicles is thus buffered, while *that inside is able to respond to pH changes*. In this case, a 10 μs light flash leads to the appearance of H$^+$ (a pH drop) inside the vesicles over an even shorter period. Thus, again proton pumping is kinetically fast—and we see that protons disappear from the outside of the vesicle and appear on the inside.

Is the proton leak slow enough?

Clearly, for protons to drive ATP synthesis, *their rate of passive transfer (leakage) across the membrane must be much less than the rate at which they are consumed in ATP synthesis*.

The rate of proton leakage can be measured, with a pH electrode, in O$_2$ pulse experiments—it is represented by the descending limb in Fig. 20.1b. Rates (when ATP synthesis is blocked with oligomycin) are at least fifty-fold less than the rate of ATP synthesis. Thus the natural rate of passive H$^+$ leakage will not significantly detract from ATP formation.

The rate of proton leakage can be increased several orders of magnitude by agents which uncouple phosphorylation (detergents, dinitrophenol, etc.). In this case, protons *are* diverted and ATP synthesis is prevented: *uncoupling agents work by increasing the permeability of membranes to protons* (see Section 34).

21 Stoichiometries in ATP synthesis

How many protons are required to make 1 ATP? The H⁺/P ratio

ATP synthesis is driven by protons moving down their electrochemical gradient. Energetic considerations set a *lower limit* on the number of protons (n) required to drive the synthesis of one ATP molecule. For example, if the proton gradient gives (for the free energy change when one proton moves)

$$\Delta\tilde{\mu}_{H^+} = -20 \,\text{kJ/mol H}^+$$

and the ATP/ADP disequilibrium gives (for the free energy change when one ATP is hydrolysed)

$$\Delta\overline{G} = -55 \,\text{kJ/mol ATP}$$

then the minimum number of protons required to synthesize one ATP molecule must be

$$\Delta\overline{G}/\Delta\tilde{\mu}_{H^+} = 2.75 \,\text{mol H}^+/\text{mol ATP}$$

Thus, for net ATP synthesis to occur

$$H^+/\text{ATP} = n \geqslant \Delta\overline{G}/\Delta\tilde{\mu}_{H^+} \;(\geqslant 2.75 \text{ in this example})$$

To avoid specifying the reaction and its direction, $\Delta\overline{G}$ for ATP hydrolysis is conveniently written as ΔG_P in subsequent discussions.

This theoretical minimum number of protons per ATP molecule will vary with conditions *in vivo*. If ATP levels are very low, the energy released on ATP hydrolysis will fall ($-\Delta G_p < 55$ kJ/mol) and this theoretical minimum will decrease. Conversely, if the proton gradient (strictly $\Delta\tilde{\mu}_{H^+}$) were smaller, more protons would be required per ATP. However, it is believed that **the actual value n— the H⁺/P ratio—is fixed in biological systems.**

This theoretical minimum ratio, as shown above, need not be a whole number. However, it is also believed (by analogy with other biochemical processes) that *the H⁺/P ratio, is a small integral number*—probably 3. In principle, the biological system sacrifices some flexibility, and efficiency, for mechanistic simplicity. A fixed stoichiometry seems to be common to all biological pumps—the stoichiometry of 3 Na⁺/ATP for the Na⁺/K⁺ pump is fixed, irrespective of the size of the existing gradient.

The fixed, integral nature of n, the H⁺/P ratio, has several consequences.

2 If ΔG_P falls (and n does not change) some free energy is dissipated as heat.

2 If $\Delta\tilde{\mu}_{H^+}$ falls, eventually there will be insufficient energy in n protons to drive ATP synthesis, and the reverse, hydrolysis, will occur.

3 It is likely that H⁺ acts by binding to an (integral) number of binding sites on the ATP synthase (see Section 27).

These consequences are all dependent on the mechanism of proton usage; thermodynamics places no such restrictions. For example, if $\Delta\tilde{\mu}_{H^+}$ falls, it would theoretically be possible to make ATP by recruiting more protons (by increasing H⁺/P); however, at present we believe that the mechanism permits only one, fixed value.

Measurement of the H⁺/P ratio

The number of protons required for the synthesis of one molecule of ATP is believed to be fixed throughout nature. This view is strengthened by similarities in the structure (and hence presumably the mechanism) of the ATP synthase of plants, animals and bacteria (see Section 25). The H⁺/P ratio is thus the fundamental stoichiometry in redox-linked phosphorylations.

Measurement of this ratio (n) is, however, not straightforward. Kinetic approaches, which effectively count H⁺ ions moved in response to a small pulse of oxygen or ATP, are simpler both conceptually and practically (see Section 22), but they are subject to various restrictions (only suitable for a few types of membrane) and artefacts (due to H⁺ leaks etc.). Thermodynamic approaches, which measure ΔG_P and $\Delta\tilde{\mu}_{H^+}$ at equilibrium balance (Section 23) are in principle more general and more rigorous—but in practice more difficult to perform.

When considering ATP synthesis in mitochondria, an additional problem arises. Mitochondria import ADP + P$_i$ from the cytoplasm, make ATP (using the ATP synthase) inside, and export the ATP. *This import/export system also uses energy from the H⁺ gradient*—in fact it requires 1 H⁺/ATP made (Section 32). Thus, attempts to measure the H⁺/P ratio in whole mitochondria will yield an *apparent* H⁺/P ratio of $n+1$, where n is as defined above—*the number of H⁺ used by the ATP synthase in ATP synthesis*.

Protons pumped by oxidation—the H⁺/O(H⁺/2e) ratio

Protons are pumped across a coupling membrane as electrons move along a chain of redox carriers. For each electron traversing a fixed redox span ΔE, the energy released is equal to $F(\Delta E)$ (Section 6). Thus, there is a *theoretical maximum* number of protons which can be pumped against a given H⁺ gradient. This is given by $F(\Delta E)/\Delta\tilde{\mu}_{H^+}$ (energy released per e transferred/energy trapped per H⁺) and clearly depends on the size of the redox span.

By analogy with the H⁺/P ratio, it is believed that, for a fixed redox span, *the H⁺/e ratio is fixed, and an integer*. It must also be smaller than the theoretical maximum (above), since some of the

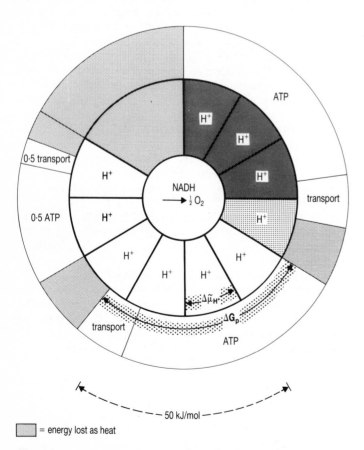

= energy lost as heat

Fig. 21.1. Relationships between H⁺/O, H⁺/P, ΔG_P and $\Delta\tilde{\mu}_{H^+}$. The angle subtended by a given sector represents available energy. Stoichiometries are indicated by number of sectors, e.g. H⁺/O given as 10, P/O $\doteqdot$ 2.5, apparent H⁺/P = 4, of which $n = 3$, and transport requires 1 H⁺ (shaded in upper right quadrant).

energy yielded on oxidation will be dissipated as heat. For historical reasons, this ratio is normally given in the form of an $H^+/2e$ ratio, where $H^+/2e = m \leqslant 2F\Delta E/\Delta\bar\mu_{H^+}$.

Oxidation of strong reductants yields more energy than oxidation of weaker ones (ΔE is greater). Organisms are able to trap this extra energy, and *m generally rises as stronger reductants are used*. In mitochondria, the $H^+/2e$ ratio when NADH is oxidized by oxygen (hence H^+/O ratio) is 10–12; when succinate is oxidized, $m = 6$–8. The measurement of H^+/O ratios, and its problems, are discussed in Section 22.

'Coupling sites'

Redox energy is coupled to ATP synthesis via a proton gradient. Thus energy is conserved when a redox reaction yields a proton and/ or absorbs a proton on the requisite side of the membrane. When a reaction between a pair of redox carriers does this, this region of the respiratory chain is said to include a **coupling site**.

The presence of a coupling site depends on the nature of the electron carriers, and is discussed in detail in Sections 24 and 25. In mitochondria, for example, oxidation of quinol by cyt b involves proton release (a coupling site); oxidation of cyt c_1 by cyt c does not. From above, there must be 5 or 6 such sites between NADH and O_2 in mitochondria ($= H^+/e$ ratio).

Note that thermodynamic considerations put a *maximum* on the number of coupling sites in a redox chain—*but not a minimum*. If the structure of the respiratory chain is such, the extra energy released when stronger reductants are oxidized may not be trapped by pumping protons and, instead, be dissipated as heat. In *E. coli*, for example, NADH and succinate are equally effective (mol ATP/mol reductant) for ATP synthesis (Section 14). We say that the coupling site(s) between NADH and quinone, found in mitochondria, are absent in *E. coli*.

Mols ATP made per mol reductant—the P/O ratio

ATP synthesis, in mitochondria, is coupled to the reduction of oxygen. The ratio between mols ATP synthesized (P) and oxygen consumed (O) is the **P/O ratio**. Dissolved oxygen can be monitored using a Clark oxygen electrode, and thus P/O ratios for mitochondria are readily measured (see Fig. 21.2). From above, we can see

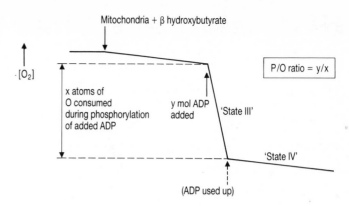

Fig. 21.2. Measuring the P/O ratio in well coupled mitochondria. Measurement of $[O_2]$ uses an oxygen electrode.

that the P/O ratio will be higher using the stronger reductant, NADH, than using succinate.

The *maximal P/O ratio*, for a given substrate (like the above ratios) is thought to be *a fixed value*. However, since the link between ATP synthesis and oxidation is indirect, and more than one proton is required per ATP made *it need not be an integral number* as was originally proposed. The most likely values in mitochondria are:

from NADH, P/O = 2.5–2.7,

and from succinate, P/O = 1.5–1.7.

The *measured P/O ratio* will vary, depending on the extent of coupling between oxidation and ATP synthesis; if the membranes are leaky to protons (uncoupled), or if the protons are used for processes other than ATP synthesis (e.g. for ion transport), the measured P/O ratio will be less than the maximal one.

A more general parameter than the P/O ratio (which relates only to systems where oxygen is reduced) is the *P/2e ratio*. In cyclic phosphorylation in chloroplasts, for example, there is no net oxidation/reduction to measure (since chlorophyll reduces itself). Without a clear chemical change, the *P/2e* ratio is very difficult to assess. A popular estimate is *P/2e* = 1.33.

22 Measurement of the H⁺/P ratio: 1 Kinetic approaches

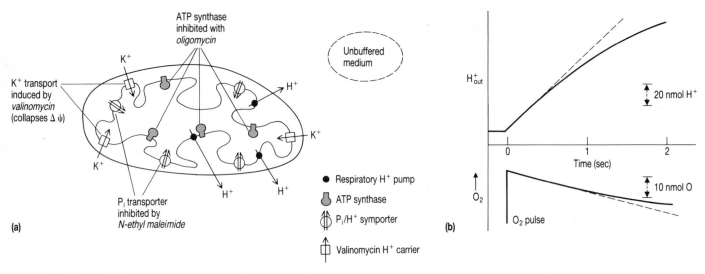

Fig. 22.1. Mitochondria optimized for an oxygen pulse experiment. (For apparatus see Fig. 20.1a.)
(b) Initial rates of O_2 uptake and H^+ extrusion by mitochondria. (Compare *time scale* with Fig. 20.1b.)

The oxygen pulse experiment in mitochondria

The paradigm for the kinetic approach to the H^+/P ratio is the **oxygen pulse** experiment, originally developed by Mitchell and Moyle. In fact, this experiment measures the H^+/O ratio in mitochondria; however, by separately measuring the P/O ratio (see Section 21), the apparent H^+/P ratio = $H^+/O \div P/O$ can be determined.

The experimental set-up is depicted in Fig. 20.1a. Typical results are shown in Fig. 20.1b. Anaerobic mitochondria incubated with substrate (succinate, β hydroxybutyrate, etc.) are given a pulse of O_2 (as H_2O_2), and extruded protons are measured with a (glass) pH electrode (ascending limb); when the oxygen is exhausted, no further extrusion occurs and the protons leak slowly back across the membrane to restore the original state (descending limb). The H^+/O ratio is calculated from nmol H^+ extruded/natom O supplied.

Quantifying the oxygen pulse experiment

So that the *observed* H^+O ratio reflects its *true* value, we must ensure that:
1 *all extruded H^+ are observed by the electrode*, and
2 *all oxygen is used up within the time of measurement.*
The latter is normally assumed, since the affinity of cytochrome oxidase for O_2 is very high—so that even the last traces of oxygen are rapidly consumed. Alternatively (see below) O_2 consumption may be measured directly with a rapidly responding oxygen electrode.

However, several factors may limit access of H^+ to the electrode:
1 *Buffering:* even in an unbuffered medium, mitochondrial proteins will buffer pH changes. This is overcome by calibrating the pH electrode in the presence of the mitochondria.
2 *Charge build-up:* H^+ movement across the membrane leads to a build-up of positive charge outside, prevention further extrusion of H^+. This is overcome by adding K^+ + valinomycin (see Section 34). K^+ movement *into* the mitochondria carries positive change, and prevents the development of a membrane potential.
3 *Use of the H^+ gradient produced:* mitochondria will tend to use H^+ for their normal purposes, e.g. ATP synthesis, ion transport. This is prevented by pretreating mitochondria with specific inhibi-

tors: *oligomycin* (to block ATP synthesis) and *N-ethyl maleimide* (to block P_i transport—the major anion transported). In the *absence of added Ca^{2+}*, no other transporter works fast enough to interfere with H^+ measurement.

These precautions are summarized in Fig. 22.1a.
4 *Passive leakage of H^+*: even if the mitochondria are prevented from using H^+, ions will *leak* down a concentration gradient. This may be overcome *either* by extrapolation of the 'leakage limb' of the pH trace (Fig. 20.1b) *or* by restricting measurement to the earlier stages of H^+ extrusion when the gradient is smaller. This latter requires more sensitive and rapidly recording pH electrodes (see below).

Initial rate measurements

With recently developed, rapidly responding electrodes for the detection of oxygen and H^+, it is possible to measure both the *rate of H^+ extrusion* and the *rate of O_2 utilization* during an oxygen pulse (i.e. during the ascending limb of Fig. 20.1b). In this case the effects of all the above, except that of buffering the protons released, are minimized (provided, of course, that the precautions given above are taken). For example, charge build-up and proton leakage both become progressively greater as more protons are moved; fewer protons moved means less 'back pressure'. Similarly, since oxygen usage is measured directly, no *assumption* needs to be made about how much is used.

Typical results are shown in Fig. 22.1b.

The consensus results from experiments of these types give (for mitochondria)

H^+/O (from NADH)   = **10–12 (H^+ per 2*e*)**

H^+/O (from succinate) = **6–8 (H^+ per 2*e*)**

Alternative measurements of pH: use of indicators

A pH indicator responds to a change in pH by changing colour. In principle, then, the pH electrode of Fig. 20.1 can be replaced by a *pH indicator* and a *spectrophotometer*. This has the advantage of a far faster response (< 1 μs) than even a rapid pH electrode (> 100 ms). A suitable indicator would have (i) its $pK_a = 6–7$, so as to respond

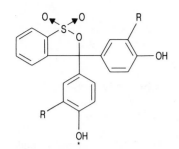

Neutral red

Phenol red (R = H)
Cresol red (see Section 20) (R = CH$_3$)
(Note increased polar nature of these impermeant indicators)

Fig. 22.2. Structures of neutral red and phenol (cresol) red indicators. * shows position of protonation/deprotonation.

maximally to changes in the physiological range; and (ii) *no tendency to bind to membranes*, which might alter the properties of either the membrane or the indicator itself.

Suitable indicators for mitochondria and chloroplasts have been found to be the permanent indicator, **neutral red**, and the *impermeant indicators*, **phenol red** and **cresol red** (Fig. 22.2). Exploitation of these indicators to measure internal and external pH in coupled vesicles, and in quantification of proton movement, is described elsewhere (Sections 20, 25).

Measurement of the P/O ratio

Determination of the P/O ratio, in mitochondria or bacteria, involves measuring oxygen uptake (using an oxygen electrode) and, over the same time, following P$_i$ incorporation into ATP (conveniently using radioactive P$_i$). Techniques are those which would be used to measure rates of other typical enzyme catalysed reactions.

However, in *well coupled mitochondria*, the phenomenon of *respiratory control* (Section 32) can be utilized. If mitochondria exhaust their supply of ADP, oxidation slows down. Thus a simple measurement of the P/O ratio can be made in mitochondria by adding a known amount of ADP, and measuring the amount of oxygen consumed *during the period of enhanced respiration* (Fig. 21.2).

The consensus results from experiments of these types (in mitochondria) are

P/O (from NADH) = 2.5–2.7

and

P/O (from succinate) = 1.5–1.7

Combining these with the values for H$^+$/O (above), these values yield an apparent H$^+$/P ratio of about 4 for mitochondria. Since the apparent H$^+$/P ratio, in mitochondria, *includes 1 H$^+$ used for transporting ATP, ADP and P$_i$* (Section 21), we have

apparent H$^+$/P = $n + 1 = 4$

i.e. *the number of protons required by the synthase for ATP synthesis* is

true H$^+$/P = $n = 3$ (H$^+$ per ATP)

The ATP pulse experiment in submitochondrial vesicles

The above procedures measure H$^+$/P indirectly, via H$^+$/O and P/O (or, in chloroplasts, P/2e) ratios. Assuming that ATP hydrolysis by the ATP synthase is the reverse of synthesis, a measurement of *the number of H$^+$ pumped per ATP hydrolysed* by the ATP synthase should give the H$^+$/P ratio directly.

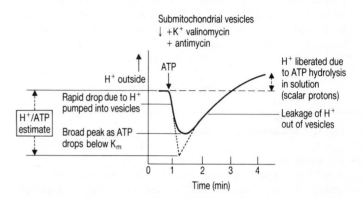

Fig. 22.3. Measuring the H$^+$/ATP ratio in submitochondrial vesicles. The apparatus is essentially that in Fig. 20.1a, except that: (i) the oxygen electrode is omitted; (ii) submitochondrial vesicles, not mitochondria, are in suspension; and (iii) ATP, not oxygen, is injected.

Thus an ATP pulse experiment can be designed analogous to the O$_2$ pulse experiment. A known amount of ATP is added to submitochondrial vesicles—which are inverted with respect to mitochondria so that their ATP synthase is on the outside—and pH followed (Fig. 22.3). K^{2+}/valinomycin (to prevent electrical potential build-up) and antimycin (to block electron flow) are added to prevent competing processes as above. (As the vesicles are inverted, ATP transport is not occurring.)

However, since K$_m$ for ATP hydrolysis is relatively high, it will take some time for all ATP to be used up, during which time leaks will become significant—and the observed ratio, typically about 2 H$^+$/ATP, is likely to be an *underestimate* of the true value.

'Equilibrium' between energy stores

If the proton gradient and ATP synthesis are *at equilibrium*, then the energy released on hydrolysis of 1 mol ATP (ΔG_P) is equal to the energy available to synthesize 1 mol ATP ($n\Delta\tilde{\mu}_{H^+}$)—when n is the apparent H⁺/P ratio. From Section 4.

$$\Delta G_P = n(\Delta\tilde{\mu}_{H^+}) = n(RT\ln[H^+]_A/[H^+]_B + F\Delta\psi) \qquad \text{(eqn 23.1)}$$

To measure n, the number of H⁺ moved per ATP made, we must:
1 Establish an equilibrium between the redox reactions and the ATP synthesis.
2 Measure ΔG_p as $\Delta G^{0'}$ is known, this involves measuring [ATP], [ADP] and [P_i]
3 Measure [H⁺] both inside and outside the vesicle
4 Measure $\Delta\psi$ across the vesicle membrane.

Equilibrium 1 is normally achieved by allowing phosphorylation (linked to electron flow) to proceed, under optimal conditions, until ΔG_P is constant. In fact, this situation represents a steady state but, in well-coupled systems, ΔG_P should approach the equilibrium value. In well-coupled mitochondria, this occurs *when oxidation slows on ADP exhaustion* ('State IV', see Fig. 21.2).

ATP and other concentrations **2** are measured, after quenching the system with perchloric acid, by chemical procedures.

ΔpH **3** and $\Delta\psi$ **4** are measured by **probe distribution methods**.

Theory of probe distribution methods

$\Delta\psi$ measurement

Vesicles which pump protons inwards (e.g. chloroplast thylakoids, submitochondrial vesicles) will acquire a membrane potential, *positive inside*. They will thus *accumulate permeant anions*. A⁻ (e.g. SCN⁻, NO₃⁻)*.

When equilibrium is reached,

$$\Delta\psi = RT/F\ln[A^-]_{in}/[A^-]_{out} \qquad \text{(eqn 23.2)}$$

and by measuring the accumulation of A⁻, $\Delta\psi$ can be measured.

In vesicles with opposite polarity—negative inside (e.g. mitochondria, bacteria)—permeant cations (e.g. K⁺/valinomycin, triphenylmethylphosphonium [TPMP⁺]) *are accumulated.*

ΔpH measurement

Vesicles which accumulate H⁺ (e.g. chloroplast thylakoids, submitochondrial vesicles) also *accumulate weak bases* (e.g. methylamine, see Fig. 23.3). The conjugate acid (RNH₃⁺) cannot cross the membrane due to its charge.

When equilibrium is reached

$$[RNH_2]_{in} = [RNH_2]_{out} \qquad \text{(eqn 23.3)}$$

since the uncharged base is permeant, and

$$K_a = \left(\frac{[RNH_2]\cdot[H^+]}{[RNH_3^+]}\right)_{in} = \left(\frac{[RNH_2]\cdot[H^+]}{[RNH_3^+]}\right)_{out} \qquad \text{(eqn 23.4)}$$

since the acid–base equilibrium is established in each compartment. Combining these equations

$$[H^+]_{in}/[RNH_3^+]_{in} = [H^+]_{out}/[RNH_3^+]_{out}$$

or

$$[H^+]_{in}/[H^+]_{out} = [RNH_3^+]_{in}/[RNH_3^+]_{out} \qquad \text{(eqn 23.5)}$$

We can measure accumulation of RNH₃⁺ only by following accumulation of total base

* Note that permeant anions are large and singly charged; Cl⁻ is too small and SO₄²⁻ too highly charged to cross a lipid bilayer.

$$[RNH_2]^{tot} = [RNH_3]^+ + [RNH_2]$$

However if pK_a of [RNH₃⁺] is high compared to pH (i.e. RNH₂ is largely protonated in both compartments)

$$[H^+]_{in}/[H^+]_{out} = [RNH_2]_{in}^{tot}/[RNH_2]_{out}^{tot} \qquad \text{(eqn 23.6)}$$

and by measuring the accumulation of base, ΔpH can be measured.

In vesicles which expel H⁺ (e.g. mitochondria, bacteria) weak acids (DMO, acetic acid) are accumulated.

Practical measurements of probe distribution

Example: measurement of $\Delta\tilde{\mu}_{H^+}$ in mitochondria

Mitochondria are incubated with β-hydroxybutrate (oxidizable substrate), ADP + P_i, and *trace amounts* of ⁸⁶Rb⁺/valinomycin, and [¹⁴C]DMO, in an oxygenated buffer at 30°C. When 'equilibrium' (State IV, see Fig. 21.2) is reached (as shown by a drop in O₂ uptake rate), the concentration of the two probes inside, or outside, the mitochondria is measured by one of a variety of methods.

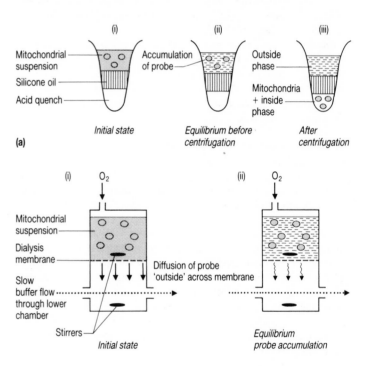

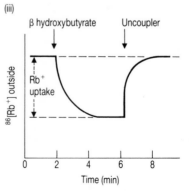

Fig. 23.1. Measurement of probe distribution in vesicles.
(a) Centrifugation method.
(b) Flow dialysis method.

1 *Filtration.* The medium is filtered through a 0.45 µm filter, by rapid suction. Mitochondria are retained by the filter and the radioactivity retained is equal to the amount of probe *inside*.

2 *Centrifugation.* Respiring mitochondria are layered onto silicone oil and centrifuged through the oil into a lower aqueous layer (Fig. 23.1a). Radioactivity in this lower layer is equal to the amount of probe *inside*.

3 *Flow dialysis.* Respiring mitochondria are incubated in a stirred chamber separated from a flow chamber by a dialysis membrane. Probe passes through the dialysis membrane at a rate proportional to its concentration *outside* the mitochondria. The concentration of probe outside can thus be deduced from the probe's rate of appearance in the flow cell (Fig. 23.1b).

4 *Ion selective electrodes.* Electrodes placed in the mitochondrial suspension can directly monitor the concentration of probe *outside* the mitochondrion. Suitable electrodes respond to $TPMP^+$, SCN^-—both of which respond to $\Delta\psi$—but not to the probes used in this example.

The *total amount of added probe* is known. Each of the above methods measures either the *concentration* of probe outside or the *amount* of probe inside; if one of these is known, the other can be calculated. However, for equations 23.2 and 23.6, we need the *concentration* of probe inside the mitochondrion; the *inside volume* must thus be measured.

Measurement of internal mitochondrial volume

To a mitochondrial suspension, incubated as above, is added [^{14}C]-sucrose (*s* cpm) (which cannot enter the mitochondrion) and [^{3}H]H_2O (*w* cpm) (which distributes into both external and internal spaces). After equilibration, mitochondria are removed (e.g. by centrifugation) and the concentration of each probe (*C* cpm/µl, *H* cpm/µl) measured in the supernatant. If *x*, *y* (µl) correspond to external and internal volumes, by simple dilution

$$C = s/x \qquad H = w/(x+y)$$

or

$$w/H - s/C = (x+y) - x = y \text{ (required internal volume)}$$

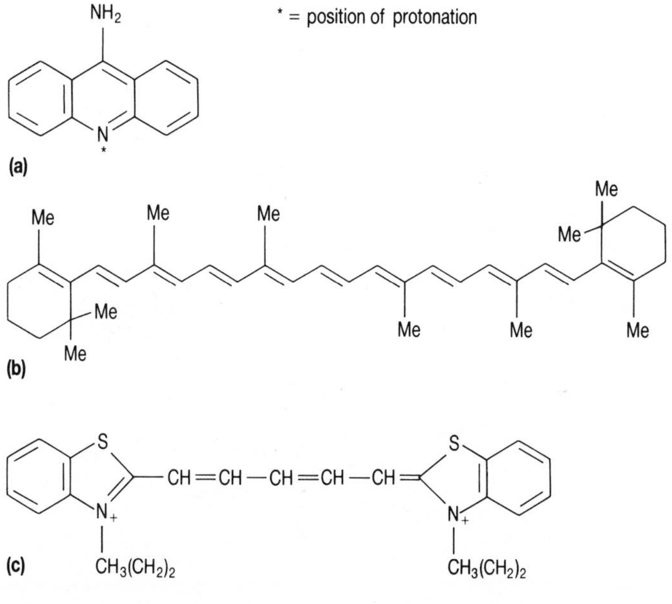

NH$_2$

* = position of protonation

(a)

(b)

(c)

Fig. 23.2. Chromophoric probes for ΔpH or $\Delta\psi$.

(a) *9-aminoacridine (ΔpH probe). (Fluorescent in dilute, unprotonated form.)*

(b) *Carotenoid ($\Delta\psi$ probe). (Note long conjugated region.)*

(c) *Carbocyanine ($\Delta\psi$ probe). (Note long conjugated region.)*

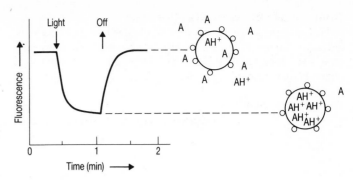

Fig. 23.3. Quenching of 9-aminoacridine fluorescence on H$^+$ uptake into chloroplast thylakoids.

Optical probes for $\Delta\psi$ measurement

Probes which distribute across membranes

The fluorescence of organic molecules is strongly influenced by their environment. The fluorescence of some fluorophores *is quenched simply by an increase in their own concentration.* This effect can be exploited if, in our probe distribution experiments, we use a weak base RNH$_2$ where R is a fluorescent group rather than a radioactive group.

A widely used probe of this type is **9-aminoacridine** (Fig. 23.2a). Like methylamine, this will *accumulate in regions of low pH*, e.g. inside chloroplast thylakoids or submitochondrial vesicles. Due to the high local concentration (and low pH), this accumulation leads to a *quenching of acridine fluorescence*, the extent of which reflects the transmembrane ΔpH (Fig. 23.3).

Probes within the membrane phase

A rather different principle is employed if an environment-sensitive probe can be fixed within the membrane itself. Such probes can be molecules with an extended π electron system, which absorb visible light. Due to a large dipole moment in the excited state, *the absorption spectrum of these molecules changes if they are oriented in an electric field* (Stark effect). The field strengths across biological membranes are sufficiently high (100 mV across a 10 nm membrane = 10^4 V/mm) for such molecules in a membrane to provide an '**optical voltmeter**' whose absorbance reflects the transmembrane potential.

Chloroplast membranes naturally contain molecules of this type, the **carotenoids** (Fig. 23.2b), whose absorbance peak (green, 550 nm) moves towards the red in energized membranes. Measurement of this intrinsic '*carotenoid shift*' thus gives a measure of transmembrane potential.

Mitochondrial, and non-photosynthetic bacteria, do not naturally contain absorbing molecules of this type. In this case lipid soluble π conjugated molecules have been synthesized and can be introduced into the membranes as *extrinsic probes*. Typical of such molecules are the **carbocyanines** (Fig. 23.2c).

Optical measurements of ΔpH and $\Delta\psi$ *are relatively fast*, since they require no separation of vesicles from medium. The carotenoid shift, for example, reflects changes in $\Delta\psi$ within 1 µs. Acridine fluorescence, which requires the probe to move across the membrane, is slower but can still reflect ΔpH changes within seconds as opposed to minutes for, e.g., the flow dialysis method.

However, with optical probes, the *quantitative relationship* between measured response (e.g. fluorescence change) and the actual change (e.g. ΔpH) *is unknown*. Care must be taken, thus, to *calibrate* optical probes by, for example, imposing artificial transmembrane potentials, in order to obtain fully quantitative information from such studies.

Consensus values of the H⁺/P ratio

Values of n, the apparent H^+/P ratio, obtained by these methods, although initially differing markedly between workers, now centre on 3–4 H^+ used per ATP made. This agrees quite well with values obtained by kinetic methods (Section 22).

We must now consider:

1 How the electron transfer chain can pump the requisite number of protons to satisfy this requirement (10–12 H^+/NADH oxidized in mitochondria); and

2 How the ATP synthase can use this number of protons to make ATP. In this regard, it is probable that the ATP synthase itself utilizes 3 H^+ to produce 1 ATP, i.e. $n = 3$; in intact mitochondria 1 H^+ is required, in addition, to transport ATP out of the mitochondrion (see Section 21), leading to a total requirement of $3 + 1 = 4H^+$/ATP made.

24 Generation of a proton gradient: 1 Mechanisms involving organic hydrogen carriers

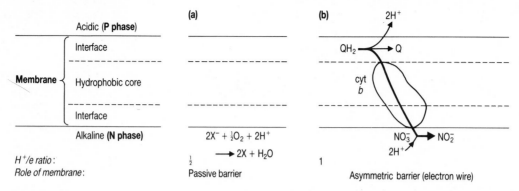

H$^+$/e ratio:
Role of membrane: | (a) | (b) |

(a) Passive barrier — $2X^- + \frac{1}{2}O_2 + 2H^+ \longrightarrow 2X + H_2O$, H$^+$/e ratio $\frac{1}{2}$

(b) Asymmetric barrier (electron wire) — H$^+$/e ratio 1

Fig. 24.1. *Above and facing:* Proton gradient formation using alternating [H] and *e* carriers. The membrane is depicted as a hydrophobic core, isolated from the aqueous phases, and two interface regions where both water-soluble and lipid-soluble entities (e.g. H$^+$ and quinone) can interact.

Note that the roles of the membrane are cumulative—in **d** for example, the membrane provides an environment to stabilize the semiquinone *in addition to the previously depicted functions.*

H$^+$/e ratios indicate apparent H$^+$ transfer, viz.

H$^+$ disappearing from alkaline side +
        H$^+$ disappearing on acid side/2 × number of electrons used

Hence if protons are simply used up on one side (Fig. 24.1a), a non-integral ratio is obtained since no corresponding protons appear on the opposite side.

(a) pH change due to redox reaction on one side of the membrane
(b) reduction of nitrate by quinol in *E. coli*. Note the use of alternate [H] and *e* carriers.
(c) oxidation of reduced flavin in *E. coli*. Note the use of a redox loop, with quinone as the mobile [H] carrier.
(d) possible redox loops in the mitochondrial respiratory chain, emphasizing the role of the Q cycle (shaded) in increasing the H$^+$/e ratio.

In cases **b** and **c**, the source of [H] for the first shown reductant is a dehydrogenase on the inside of the membrane. This is omitted in these diagrams for clarity, since it is not a proton generating reaction. However it is shown in **d** for completion.

———— path of electron down electron transfer chain
—·—·— path of recycled electron in the Q cycle
–––––– path of oxidized Q
Curly lines show the flipping of quinol/quinone across the membrane.

Protons and redox reactions

From thermodynamic arguments, *any* gradient can serve as an energy store. Furthermore, the amount of energy stored in a gradient of monovalent cations is *irrespective of the particular ion involved* (see Section 20).

Why then do biological systems employ *proton* gradients as a temporary energy store? Probably because redox reactions are used as a primary energy source, and *protons participate directly in redox reactions.*

In chemical terms, oxidation is defined as *removal of electrons (e) or of hydrogen atoms ([H]) from a compound.* These two processes, which are mechanistically distinct, become equivalent only when protons are taken into account. Thus, both the following half reactions are oxidations:

$$^-OOC \cdot CH_2 \cdot CH_2 \cdot COO^- \rightarrow {}^-OOC \cdot CH:CH \cdot COO^- + 2[H]$$
    succinate   fumarate

$$Fe^{II} \rightarrow Fe^{III} + e^-$$
    ferrous   ferric

Coupling the two together produces a redox reaction which generates protons in solution.

$$^-OOC \cdot CH_2 \cdot CH_2 \cdot COO^- + 2Fe^{III} \rightarrow$$
$$^-OOC \cdot CH:CH \cdot COO^- + 2Fe^{II} + 2H^+$$

If such a reaction takes place inside a biological compartment (e.g. within a membrane-bounded organelle) the pH inside the compartment will drop (Fig. 24.1a). Mechanisms for generating a proton gradient discussed in this section utilize this property of redox reactions.

Alternating hydrogen and electron carriers

Oxidation of a [H] carrier (like succinate) by an *e* carrier (such as a cytochrome) will liberate protons. Conversely, oxidation of an *e* carrier by a [H] carrier will take up protons. If these reactions are arranged to occur on *opposite sides of a membrane*—by correctly positioning the enzyme active sites—protons will disappear from one side of the membrane and appear on the other.

As an example, Fig. 24.1b shows the nitrate reductase of *E. coli.* As electrons pass from quinol (outside) to nitrate (inside), via cyt *b* (Section 14), protons appear outside the cell and disappear inside—a pH gradient is set up. Fig. 24.1b also shows the *stoichiometry* of the process; 2 H$^+$ disappear inside, and 2 H$^+$ appear outside, for 2 *e* transferred, i.e. the H$^+$/e ratio is 1.

In this example, the H$^+$/e ratio is seen to be fixed *by the nature of the reacting species*, rather than by the energy available in the span of redox potential. The same H$^+$/e ratio would be observed if, say, β-hydroxybutyrate (outside) were oxidized by oxygen (inside) despite ΔG^0 for this latter reaction being some threefold greater. This has led to the evolution of mechanisms to amplify this ratio.

Redox loops

To increase the yield of proton gradient from a given reductant, Mitchell suggested that electrons *traverse the membrane several times* on their way to oxygen. In one direction they cross via *e* carriers (cytochromes, iron–sulphur proteins), and in the other direction via [H] carriers (flavin, quinone)—hence 'loops'. This gives rise to repeated proton uptake at one side of the membrane and release at the other. *The H$^+$/e ratio rises to equal the number of loops traversed.*

A two loop system is shown in Fig. 24.1c, FMNH$_2$, on the *outside* of the membrane, is oxidized by iron–sulphur proteins which in turn reduce quinone (Q) on the *inside*. The hydrogen atoms from quinone are used to reduce cyt *b* on the outside and the electrons carried, via more cytochromes, to oxygen *inside* the membrane. Such a simple two loop system may occur in *E. coli* when the terminal electron

mechanisms described in this section, in that the protons taken up and released come from *organic cofactors* in the respiratory chain.

The other mechanism, based on *proteins pumping H*$^+$ across the

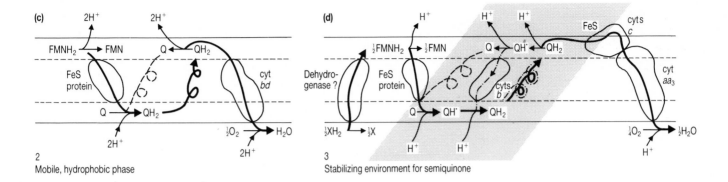

carrier is cyt *bd* (see Section 14).

In most of these reactions, *e* or [H] are passed across the lipid membrane *between redox centres fixed relative to each other*, in the respiratory complexes (see Section 12). The quinones are the exception to this generalization; both quinone and its reduced form *can flip from one side of the membrane to the other*, carrying [H] directly across the membrane (curly arrows).

Note that most dehydrogenases which produce reduced flavin in such loops, are located with their active site on the inside of the membrane, e.g. NADH dehydrogenase (Section 12). This requires an extra limb to the first loop, to carry [H] from the site of donation to the outside of the membrane, where the first protons must be generated. The glycerol-1-phosphate dehydrogenase in mitochondria is an exception to this general rule in facing outwards, towards the cytoplasm.

Evidence for redox loops

There is no direct evidence for the existence of redox loops beyond a general assumption that, since H$^+$ ions *are* released in redox reactions, these protons will contribute to any proton gradient observed. However, supporting evidence is given by the following

1 Both [H] and *e* carriers participate in electron transport, and are arranged asymmetrically in the membrane (Sections 11–18). This is necessary for the operation of redox loops—but is not proof of their existence.

2 Ubiquinone and ubiquinol, despite their size, can 'flip' across (at least artificial) membranes. This again is required by the redox loop model (see Fig. 24.1c).

Limitation of redox loop mechanisms

We have seen, above, that substrates for the respiratory chain are generally [H] donors (succinate, lactate, β-hydroxybutyrate etc.). Within the respiratory chain, however, there are only *two [H] carriers, flavin and quinone*. Thus the maximum number of redox loops possible—and the maximum H$^+$/*e* ratio they can generate—is limited to two. This restriction would lead to a loss of energy released on oxidizing highly reducing substrates (e.g. NADH).

Thus two additional mechanisms exist to increase this ratio further. **The Q cycle** is based on the ability of a quinone to form a *semiquinone radical* by the addition of one electron. Such a radical can be stable (only) when bound to protein. The mechanism of gradient generation by the Q cycle is formally equivalent to the

lipid bilayer, is different in principle and is dealt with in the next section.

The Q cycle

In the Q cycle, as one electron is released from the FeS centre of a dehydrogenase, it reduces quinone to a semiquinone radical (QH$^{\cdot}$). This radical is inherently very unstable, and must be *stabilized* within the membrane by being *bound to a protein*. As a consequence, the radical (and thus the electron) cannot traverse the lipid bilayer.

For transfer across the membrane, this electron must be accompanied by a second electron, so that the semiquinone is reduced to the mobile quinol. As above, transfer of the quinol across the membrane leads to the disappearance of 2 H$^+$ on the side of reduction, and the liberation of 2 H$^+$ on the side of oxidation. In the Q cycle, however, when the quinol is reoxidized, only 1 *e* passes on down the chain to cyt *c* and oxygen; the second is *recycled*, via the two *b* cytochromes, to accompany subsequent electrons across the membrane. Thus *two protons cross the membrane while only one electron is 'used up'*— the second electron can be considered to play a 'catalytic' role. As can be seen from Fig. 24.1(d), this further increases the H$^+$/*e* ratio, in the mitochondrial respiratory chain, to 3. This is probably the limit achievable by 'redox loop' mechanisms; further increase to the measured value of H$^+$/*e* = 5–6 (Section 23) involves H$^+$ pumping by proteins (Section 25).

The Q cycle retains the concept of proton generation and uptake by alternating organic [H] carriers and inorganic *e* carriers. It is novel in that it places the *b* cytochromes on a side path in the respiratory chain, and that *the semiquinone radical is both oxidized and reduced by b cytochrome*. This is possible only because this radical is unstable—it shows a high tendency either to accept or to donate electrons to reach the more stable quinol or quinone.

Evidence for the Q cycle

1 If electron transfer is blocked after cyt *b* by antimycin, and *oxygen* introduced, cyt b_{566} becommes more *reduced*! This is because oxidation of QH$_2$, via cyts *c* and *a*, produces QH$^{\cdot}$, which in this scheme is the reductant of cyt *b*.

2 EPR signals corresponding to two different (semiquinone) radicals have been detected in this region of the electron transfer chain in mitochondria. Presumably one represents QH$^{\cdot}_{out}$ and one QH$^{\cdot}_{in}$.

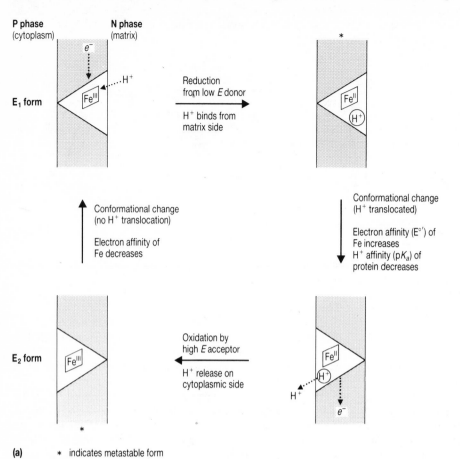

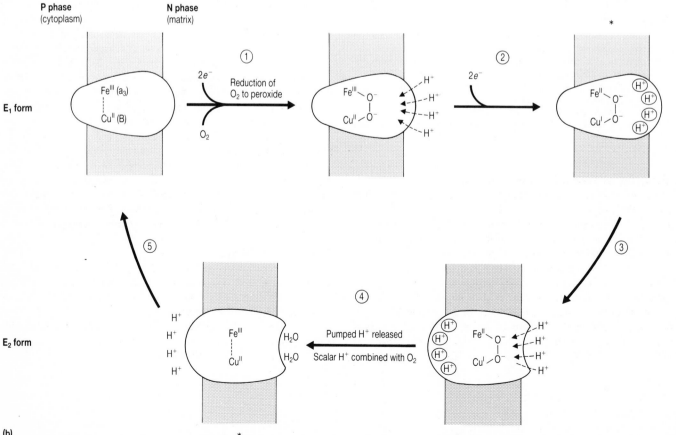

Fig. 25.1. **(a)** The general scheme for a redox dependent pump.

(b) Scheme for H^+ pumping by cytochrome oxidase. (Note that the redox stages are simplified relative to Fig. 13.1 to emphasize the stages of H^+ uptake/release.)

Redox linked proton pumps

As outlined in the previous section, alternating [H] carriers and *e* carriers can build up a transmembrane proton gradient across a membrane. *Electrons and protons cross the membrane together, as [H]* (Fig. 24.1). As noted in Section 24, this 'loop' mechanism could lead to energy losses, because in it the H^+/e ratio is dependent on the number of organic [H] carriers rather than the available energy, ΔG ($= -F\Delta E$).

To approach maximum efficiency, mitochondria augment 'looped' proton translocation with **redox-linked proton pumps**. These pumps actually *move protons across the membrane without combining H^+ with e^-* (Fig. 25.1b), much as the sodium pump moves Na^+ ions across a membrane. In a redox linked pump, however, *the energy source for ion movement is the concomitant oxidation*.

Mechanism of H⁺ translocation

A protein (E) carrying out redox-linked proton translocation must (i) *alter its proton affinity (pK_a) on oxidation and reduction* and (ii) be arranged so as to *take up and release H^+ on opposite sites of the membrane*. This is summarized in the scheme

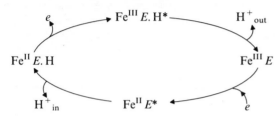

with the starred species being unstable at the ambient pH. It is assumed that the (more negative) reduced species has the higher affinity (higher pK_a) for the (positively charged) proton. The energetics of the system can be readily understood in terms of the formalism given in Section 44 for the Na^+/K^+ pump, and the required conformational changes in terms of Fig. 25.1a.

This change in proton affinity on oxidation/reduction is analogous to the Bohr effect in haemoglobin—which releases H^+ on oxygenation. Indeed, the uptake/release of H^+ by the transmembrane electron transfer complexes may be termed a 'membrane Bohr effect'. This is a useful analogy, since it emphasizes the linked nature of oxidation and H^+ release. In haemoglobin, for example, since oxygenation causes H^+ release, a drop in pH (raised H^+) promotes oxygen release. Similarly, in redox carriers, if redox state affects pK_a (above), then *pH will affect redox potential*. ΔpH dependence of $E^{0'}$ for a redox carrier can be used as a diagnostic test for a proton pumping site in the respiratory chain (see Section 6).

Also, by analogy with haemoglobin, the amino acid residues whose H^+ affinities change during oxidation/reduction *may be distant from those interacting with the Fe (or Cu) cofactors*, i.e. the pK_a changes are probably mediated via conformational changes within the protein's tertiary structure. As a result, we are unlikely to know which particular acid side chains change their ionization until we know the detailed protein structure.

Structural aspects of a cytochrome aa₃ proton pump

In the mitochondrial respiratory chain, the best characterized proton pump is cytochrome oxidase (complex IV). On reducing oxygen (a [H] carrier) with electrons from cyt *c* (an *e* carrier) via the oxidase, 1 H^+ must disappear from the matrix as $O_2 \rightarrow H_2O$. It is found, however, that for each electron passing through the oxidase, *a second proton disappears from the mitochondrial matrix and appears outside*. This proton is transferred from one side of the membrane to the other via the protein molecule, i.e. it is pumped across. Evidence for this proton stoichiometry is given below.

The oxidase shows the expected features of a proton pump.

1 It spans the membrane and is in contact with the aqueous phase on both sides.

2 Reduction of the oxidase causes conformational changes in the protein (as shown by changes in protein fluorescence or reactivity with cyanide).

3 The standard redox potentials (E^0) between the intermediates in oxidation (see Section 13) are pH dependent. Detailed analysis of the effects of pH on E^0 indicates that *H^+ pumping is associated with the final stages of oxygen reduction, from peroxide (O_2^{2-}) to water (H_2O)*. Reduction of O_2 to O_2^{2-} is not associated with H^+ movement.

These features are incorporated into Fig. 25.1b, where the general scheme for a redox pump (Fig. 25.1a) is expanded to describe cytochrome oxidase. Like all pumps, cytochrome oxidase must exist in two conformations, E_1 and E_2, whose ion binding sites have access to opposite sides of the membrane. The energy required to lower the binding affinity for H^+, for release on the low pH (P) side of the membrane, comes from the reduction of peroxide to H_2O.

(In the figure, the haem a/Cu_A centres are omitted as they do not participate directly in H^+ movement.)

1 Haem a_3 and Cu_B are reduced by 2 *e*. O_2 bound to the binuclear centre is reduced to O_2^{2-}, and 4 H^+ bind to the protein from the matrix side.

2 Two further electrons enter the a_3/Cu_B centre.

3 The enzyme changes conformation from $E_1 \rightarrow E_2$. The bound protons are now exposed to the cytoplasmic side of the membrane, and less tightly bound to the protein.

4 Electrons pass to O_2^{2-}, forming water (with the uptake of 4 'scalar' protons). The 4 'pumped' protons are released from their binding sites into the aqueous P phase (cytoplasmic side). Thus a net 8 H^+ is lost from the matrix (N phase) and 4 H^+ gained by the P phase per 4 *e* transferred to O_2.

5 The association between haem a_3 and Cu_B is restored, and the fully oxidized enzyme relaxes back to the E_1 conformation.

The nature of the proton binding groups ('proton channel') remains uncertain. It has been speculated, by analogy with the ATP synthase (see Section 30), that the channel may involve *intramembrane carboxylic acid (–COOH) groups of the protein*. In particular, *subunit III of cytochrome oxidase* may be involved with H^+ transfer. This subunit contains no Fe or Cu, and, indeed, without it electron transfer through cytochrome oxidase proceeds unaffected. However, removal of this subunit, or its chemical modification by dicyclohexylcarbodimide (DCCD, which specifically modifies intrinsic membrane carboxylic acids) leads to a severe decrease in proton pumping ability (to about 30 per cent of normal). This suggests a possible—if still ill-defined—role of subunit III in proton translocation by cytochrome oxidase.

Other redox-linked H⁺ pumps

Demonstrating that cytochrome oxidase pumps protons directly (see below) was facilitated (i) by the absence of organic [H] carriers from this region of the respiratory chain (ensuring no 'loop' mechanism can operate), (ii) by the ease of isolation and *oriented reconstitution of pure complex IV into liposomes* (ensuring no interference from other regions of the chain), and (iii) by the ability to separate electron transfer reactions (involving subunits I and II) from H^+ transfer (requiring an intact subunit III). The cytochrome *bo* complex of bacteria (see Section 14) also acts as a proton pump, in an analogous manner.

Study of the other electron transfer complexes is more difficult, because they contain [H] carriers (flavins, quinones) and thus *can* transfer H^+ using loop mechanisms. In yeast, NADH dehydrogenase may or may not pump protons depending on its possession of a particular FeS centre—this is suggestive of a redox-linked pump since [H] carriers are present in both strains. There is also some evidence for *DCCD sensitivity* and/or *pH dependent redox*

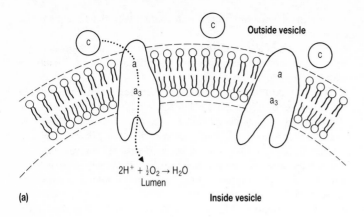

Outside vesicle

$$2H^+ + \tfrac{1}{2}O_2 \rightarrow H_2O$$
Lumen

(a)

Inside vesicle

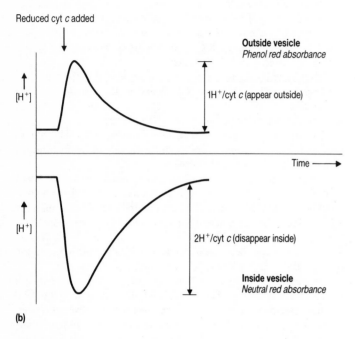

Reduced cyt *c* added

Outside vesicle
Phenol red absorbance

[H⁺]

1H⁺/cyt *c* (appear outside)

Time

[H⁺]

2H⁺/cyt *c* (disappear inside)

Inside vesicle
Neutral red absorbance

(b)

Fig. 25.2. **(a)** Cytochrome oxidase incorporated into liposomes in oriented manner.

(b) Measurement of H⁺ inside and outside liposomes. Note that for every cyt *c* oxidized, 1 H⁺ appears outside the liposome and two disappear inside.

potentials in both complexes I and III—suggestive (but not conclusive) of pump mechanisms. Finally, the existence of H⁺ pumping has been suggested from overall measurements of H⁺ stoichiometries of electron transfer (Section 23). Further evidence, however, is needed to establish both the existence and mechanism of H⁺ pumping in these other respiratory complexes.

Demonstration of direct H⁺ pumping by cytochrome oxidase

To investigate the effects of electron flow through cytochrome oxidase alone, the oxidase is isolated pure and reconstituted into liposomes. Electrons are supplied by adding reduced cyt *c*, and oxygen is ultimately reduced to H₂O (Fig. 25.2a).

To investigate H⁺ pumping, we need to measure [H⁺] simultaneously on both sides of the liposomal membrane. We use indicators—impermeant phenol red to measure external pH, and permeant neutral red (in the presence of an impermeant buffer externally) to measure internal pH (see Section 20). K⁺/valinomycin is also added, to allow H⁺ movement unhindered by a growing membrane potential (see Section 22).

Fig. 25.2b shows that, as cyt *c* is oxidized, protons appear outside the vesicle. Inside the vesicle, protons disappear. The observed stoichiometry indicates that *per electron transferred, 1 H⁺ appears on the outside of the vesicle while 2 H⁺ disappear inside—i.e. 1 H⁺ is pumped across the membrane* and one 'scalar' proton is combined with O₂ in the reaction

$$O_2 + 4e^- + 4H^+ \rightarrow 2H_2O$$

26 The ATP synthase complex

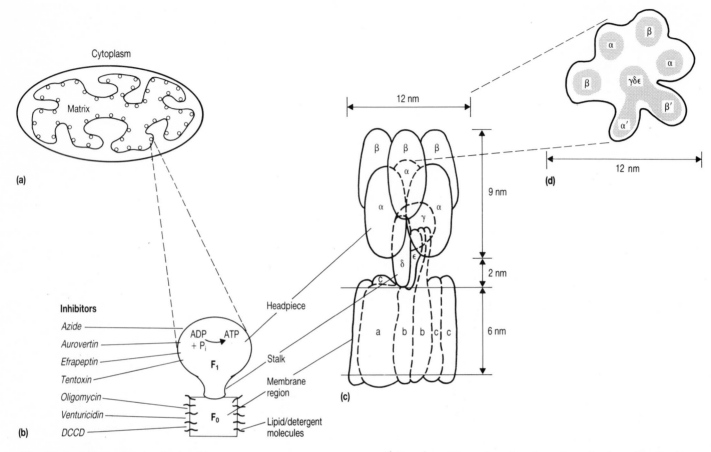

Fig. 26.1. ATP synthase in mitochondria.

(a) F_1 observed as spheres on inner mitochondrial membrane (in negative stained electron micrographs).

(b) Detergent-extracted F_1F_o ATP synthase (Complex V).

(c) Side view of ATP synthase, showing positions of subunits.

(d) Top view of F_1 portion of synthase. One pair of α and β subunits (designated α' β') is associated with central mass (γ δ ϵ subunits), inducing asymmetry.

(c and d are derived from electron micrographs using image reconstruction techniques.)

Universality of energy trapping

Energy yielded in redox reactions is stored, temporarily and locally, as a transmembrane gradient of protons. ATP synthesis, the final stage in the energy conservation system, must draw upon this transmembrane gradient for energy. This requires a *transmembrane enzyme complex*, the **ATP synthase**, which transduces gradient energy into anhydride bonds.

We have seen how, despite recurring *patterns* in electron transfer, redox carriers vary considerably between mitochondria, chloroplasts and bacteria. Indeed, the same reaction may be catalysed by different carriers in different organisms (e.g. the oxidases, cyt aa_3, cyt bo and cyt d). The ATP synthase, in contrast, shows a remarkable similarity of structure in all organisms, indicating (i) a fundamental role in all metabolism and (ii) an early evolutionary origin. The following description relates to the ATP synthase of both prokaryotic and eukaryotic organisms.

F_1 and F_o

The ATP synthase can be observed by electron microscopy as 8 nm spheres on the surface of (negatively stained) coupling membranes—for example the mitochondrion (Fig. 26.1a). Rather surprisingly, the extramembrane portion of the ATP synthase is not covalently attached to the intramembrane section and can be removed by relatively mild treatments (e.g. incubation at low ionic strength).

The soluble fragment, known as F_1 (which corresponds to the 8 nm spheres) can hydrolyse ATP, and is hence thought to contain *the active site for ATP synthesis*. (Note that, in the absence of energy inputs hydrolysis occurs as the reverse of ATP synthesis.)

After F_1 removal, the intrinsic membrane sector of the ATP synthase that remains is F_o. This traverses the membrane and must *transport H^+ from the high concentration side of the membrane to F_1*, for use in ATP synthesis. Each F_o peptide contains at least one transmembrane α-helix. F_o can be removed, like other intrinsic proteins, only if the membrane is disrupted with detergent (e.g. cholate).

If the intact coupling membrane is treated with mild detergents, F_1 and F_o copurify (to a colourless protein—no iron or prosthetic groups) as the entire ATP synthase F_1F_o complex (Fig. 26.1b). By analogy with the four major electron transfer complexes (Section 13), the mitochondrial ATP synthase has been designated complex V.

The minimal structure of F_1F_o comprises 8 different polypeptides, 5 of which are in the F_1 fragment. In higher organisms, 3–4 additional peptides may be present, but these are not directly involved in the ATP synthesis mechanism. Instead, they are probably involved in regulation of the synthase. Image reconstruction methods show F_1 as roughly *hexagonal*, with three pairs of $\alpha\beta$ dimers around a central well (Fig. 26.1c). This is occupied by the remaining 3 polypeptides, $\gamma\delta\epsilon$, (present as one copy each), which are oriented asymmetrically, contacting only one of the 3 αβ pairs (Fig. 26.1d). F_1 is attached to F_o via a slender stalk, which consists partly of the extramembrane region of the b subunits of F_o and partly of the smallest F_1 subunits, δ and ε (Fig. 26.1c). The overall stoichiometry is probably $[\alpha_3\beta_3\gamma\delta\epsilon]$ $[ab_2c_{12}]$ (Greek symbols referring to the F_1 subunits), although the precise number of c subunits is unknown.

Reconstitution studies

If F_1 is (gently) removed from energy conserving membranes (e.g. by washing at very low ionic strength), ATP synthesis is abolished, even though electron transfer can still proceed. Electron transfer is *uncoupled* from ATP synthesis. Rebinding F_1 restores synthesis; F_1 may be termed a **coupling factor**.

F_o and the F_1F_o complex can be removed from membranes only with detergent—and they will not recouple depleted membranes, which are necessarily fragmented. However, by combination with phospholipid (by sonication, or by slowly removing detergent from a detergent/lipid/protein mixture), they can be incorporated into liposomes, and their transport properties studied. It is found that F_o alone increases proton permeability in liposomes, while the incorporated F_1F_o complex will make ATP under the influence of an (induced) pH gradient. Such experiments confirm a *functional separation of properties* between F_1 and F_o to match their *structural partition*. This contrasts with the ion pumps of the plasma membrane, where ATP hydrolysis and ionophoric activity share a single peptide (Section 43).

Inhibitor studies

Inhibitors of the ATP synthase have been used to investigate its function, much as have electron transfer inhibitors in studying the electron transfer chain. ATP synthase inhibitors fall into two categories (see Fig. 26.1b). F_1 **inhibitors** bind directly to F_1, in most cases to its catalytic β subunits, preventing turnover both on and off the membrane. They may be simple ions like azide (N_3^-), complex organic molecules (the *aurovertins*), or even peptides (*efrapeptin*, *tentoxin*). These compounds are highly toxic (azide is an effective bacteriocide) as they prevent the bulk of ATP synthesis in cells. Typically (excepting tentoxin), they act throughout the animal, plant and bacterial kingdoms, showing a remarkable conservation of F_1 structure throughout evolution.

A second class of inhibitors are the **energy transfer inhibitors**. Examples are *oligomycin, venturicidin* and *dicyclohexyl carbodiimide* (DCCD). These bind to F_o, blocking proton transfer, and they thus inhibit ATP synthesis. Interestingly, these compounds also block ATP hydrolysis by F_1, when it is assembled into an F_1F_o complex. This demonstrates a *tight, functional coupling between F_1 and F_o*; hydrolysis on F_1 is prevented if the obligatory pathway for proton removal, through F_o, is blocked. If the F_1–F_o link is physically broken, and when F_1 alone is isolated, this coupling is disrupted and energy transfer inhibitors no longer inhibit hydrolysis by F_1.

This coupling must involve protein/protein interactions between F_1 and F_o—presumably in the stalk linking the two. The precise

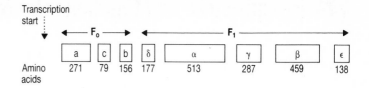

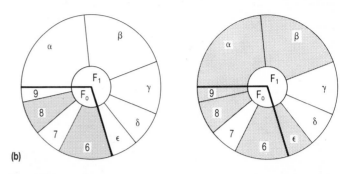

Fig. 26.2. (a) Order of genes in *atp* operon from *E. coli*. Numbers indicate amino acid residues in each subunit.

(b) Origin of F_1 F_o subunits; *left*, human mitochondria; *right*, spinach chloroplasts. Shading indicates coding on organelle genome.

nature of the interactions is uncertain; however a general model would require the F_o component to 'brake' F_1 turnover until it was suitably protonated. In isolated F_1, this brake is removed and water is protonated in its place (see Section 30).

Genes for the ATP synthase complex

In *E. coli* the 8 peptides of the ATP synthase complex are transcribed from a single operon (the *atp* operon) of about 7000 base pairs. The order of the genes is as shown in Fig. 26.2a. Mutation in any of these genes can lead to a strain able to respire, but unable to make ATP through respiration—an uncoupled mutant. (The original name for this operon was the *unc* operon.) In these mutants, substrate level phosphorylation is still possible, i.e. the mutants can live on fermentable substrates like glucose, but not non-fermentable ones like succinate.

In other bacteria, genes for the ATP synthase may be split—for example, in *Rhodospirales*, the 3 F_o genes comprise one operon, and the 5 F_1 genes a second. In eukaryotes, the situation is even more complex; some of the peptides are coded in the organelle genome (mitochondrial or chloroplast), while others are coded in the nucleus (Fig. 26.2b). Close collaboration between the two genomes is required for assembly of a fully functional ATP synthase complex.

Knowledge of the sequences of these genes has been vital in increasing our understanding of the ATP synthase mechanism (Section 27).

27 F₁—structure and function relationships

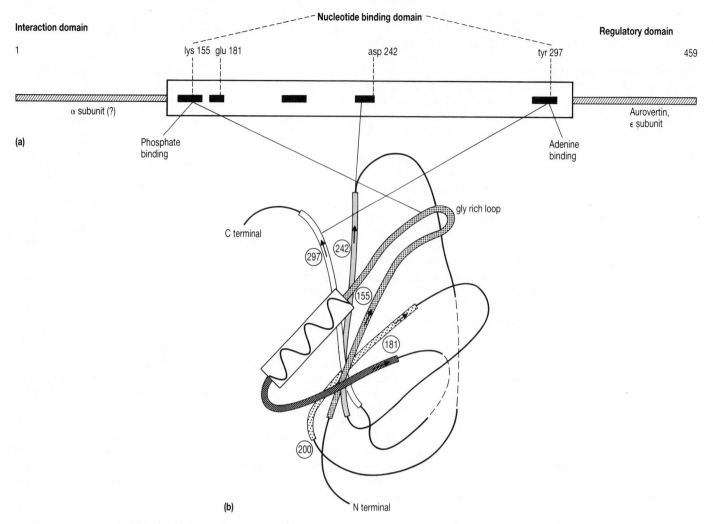

Fig. 27.1 β subunit of F₁ and its folding pattern.
(a) Domain structure of β subunit. Nucleotide binding region shown twice scale, with critical residues marked. Five β strands are indicated by bars.

(b) Possible folding pattern of ATP binding domain (note 5-strand parallel sheet).

The active site of F₁ is on its β subunit

The ATP/ADP binding region of the F₁Fₒ ATP synthase lies on the F₁ moiety, outside the membrane phase (see Section 26). However, this fragment of the synthase is itself a large protein (mol. wt. 350–400 kDa). ATP must thus interact over only a small fraction of the protein surface, its *active site*.

It is no easy task to identify the active site residues within this protein. Location of the active site of a redox protein is relatively simple—it is marked by a bound electron transfer cofactor such as haem or an iron–sulphur cluster, and can frequently be identified from the polypeptide sequence (see Section 10). However, a site of ATP binding in a protein is simply a particular three-dimensional array of amino acid side chains. It is thus difficult to identify such a site by inspection.

Attempts to identify the active site of F₁ include:

1 Locating the binding sites of covalent inhibitors of F₁ (Nbf-Cl, DCCD) which inhibit at a 1 : 1 ratio (*chemical labelling*).
2 Locating the binding sites of ATP analogues which have been modified to bind covalently (2 azido ATP and fluorosulphonyl benzoyl adenosine, see Fig. 28.3) (*affinity labelling*).
3 Inspecting the primary sequence of the subunits of F₁ for *motifs*

which are reminiscent of ATP binding sites in other proteins.
4 Study of the nature of the defects (e.g. lack of nucleotide binding etc.) in mutants in which the function of F₁ has been altered. All four approaches identify the second largest subunit (β) of F₁ as bearing the active site. This is confirmed by studies on purified β subunits of F₁ from various species; they can be shown to bind ATP directly and, albeit slowly, hydrolyse it.

Note that each molecule of F₁ is an α₃β₃γδε assembly (Fig. 26.1) and so must bear 3 catalytic sites. This has implications for the mechanism of catalysis, as is discussed in Section 29.

In the following discussion, the principles relate to F₁s from all species. However, the numbers refer to positions in the amino acid sequence of *E. coli* F₁.

Structure of the active site

The primary structure of a typical F₁-β subunit is shown schematically in Fig. 27.1a. In fact, the amino acid sequence is very strongly conserved between species (> 70 per cent identity between *E. coli* and cow) as befits the catalytic core of this enzyme. Labelling studies indicate that the ATP binding site occupies the central region of this sequence, affinity labels in particular clustering in the region 280–

300. The central region of the β subunit of F_1 may fold into an **ATP binding domain**, leaving the C and N terminals to *interact with the other subunits of F_1*.

Within the ATP binding domain are two motifs, gly.X.X.gly.X.gly.**lys**.thr (residues 149–156), and leu.leu.phe.ile.**asp** (residues 237–242) (residues in bold identified in Fig. 27.1a) which are associated with the ATP binding region of a number of kinases (from adenylate kinase to the protein kinases of oncogenes). It is suggested that the ATP binding domain of F_1 structurally resembles that of kinases (and myosin) which show these motifs. (The P-type ATPases, Section 43 do not.) Since the crystal structure of a kinase, adenylate kinase, is known, it is possible (by a judicious mixture of structure prediction and guesswork) to tentatively model the three-dimensional arrangement at the active site. This is shown in Fig. 27.1b. The glycine rich sequence, above, is seen to form a rigid loop in the centre of this region, with the remainder of the binding site comprising a 5-strand parallel β-sheet.

This model does not provide enough detail to orient the ATP molecule precisely, nor to identify specific side chains interacting with it, although it is likely that the second (β) phosphate of ATP interacts with the loop residue **lys** 155 which labels with Nbf-Cl, and the adenosine with residues around 280–300, which label with affinity ATP analogues (Fig. 28.3). One important residue in catalysis is glu 181; when this is modified, either chemically with DCCD or by mutation, activity of F_2 falls to 1% of normal. However, it is likely that catalysis by F_1 involves the collaboration of several side chains in this catalytic domain.

Roles of the remaining F_1 subunits

Although the catalytic sites reside on the 3β subunits of F_1, the remaining subunits are no less critical to its functioning. Dissociation/reconstitution experiments indicate the high rates of ATP hydrolysis require complexes of the three largest subunits α, β and γ (probably in the proportions $\alpha_3\beta_3\gamma$). ATP synthesis requires a full complement of all 5 subunits.

The following functions have been suggested for the α, γ, δ and ε subunits of F_1. Note, however, that allocation of a particular function to a particular subunit may be too simple in a system where all collaborate in producing an integrated whole.

ε subunit

The ε subunit has two major roles. First, it is necessary for F_1 to bind *correctly* to the membrane (F_o). In the absence of ε, F_1 binds to membranes containing F_o but cannot catalyse coupled ATP synthesis. This is consistent with its position in the 'stalk' (Fig. 26.1). This activity seems to involve the N-terminal region (residues 1–75) of ε, as shown by genetically truncating this subunit (Fig. 27.2a).

The ε subunit of bacteria and chloroplasts will inhibit the activity of F_1. This is taken as an indication that this subunit may have a **regulatory role** *in vivo*, perhaps in inhibiting ATP hydrolysis by non-energized F_1 molecules. This function requires the central region of the subunit sequence (residues 80–90), and appears to be taken over by a separate regulatory subunit in mitochondria (Section 33). (Thus mitochondrial F_1 contains a truncated ε subunit and an additional, inhibitory subunit, IF_1.) Inhibition involves the interaction of this sequence with the C-terminal region of the β subunit (see Fig. 27.1a).

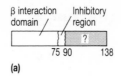

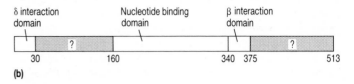

Fig. 27.2. Functional regions of other F_1 subunits.
(a) ε subunit.
(b) α subunit.

δ subunit

The major role of the δ subunit is in binding F_1 to the membrane (as, again, indicated by reconstitution studies). It is elongated and highly helical, interacts with the b subunit of F_o, and contributes largely to the 'stalk' of the complex. It may also interact with the N-terminal of the α subunit (Fig. 27.2b).

γ subunit

The γ subunit of F_1 can be looked upon as an 'organizing centre'; without it the 3α and 3β subunits do not form an active ATPase. This subunit interacts with one αβ pair in the trimeric complex, inducing an asymmetrical organization (Fig. 26.1) and hence cooperativity during turnover (Section 29).

The γ subunit has also been implicated in the proton gate, the mechanism for 'counting' H^+ ions as they cross the membrane. This ensures that the synthesis of each ATP molecule is coupled to the transfer of 3 H^+ across F_0. Damage to the γ subunit allows H^+ to pass without ATP synthesis—the membranes become 'leaky' and uncoupled.

α subunit

The α subunits form over 50 per cent of F_1 by mass, and yet their function remains a mystery. They are essential for activity; a single point mutation in α can lead to almost total inactivation, despite the fact that F_1 assembles perfectly normally. Even more mysteriously, the α subunit shows significant sequence homology with the β subunit, even down to the possession of an ATP binding site in its central domain (Fig. 27.2b). However, this binding site does not turn over during enzyme activity—the ATP present simply remains bound while substrate is bound to, and product released from, the catalytic site. The function of this 'tightly bound' ATP— like that of the α subunit itself—is unclear.

Analysis of mutations and proteolytically nicked α suggest that the N-terminal region (residues 1–30) interacts with the δ subunit, and is thus needed for membrane attachment. The region of α just on the C-terminal side of the nucleotide binding site is involved in αβ cooperativity; mutants in this region abolish cooperative turnover; and hence nearly all observable activity. Whether this implies a direct interaction of this region with the β, and/or γ, subunits is unknown.

28 Mechanism of ATP synthesis: 1 Energetics

Effect of enzyme binding on the energy levels of reaction intermediates

Synthesis of ATP, *in vivo*, requires 55–60 kJ/mol (see Section 7). In a one-step reaction, this energy would have to be supplied chemically to dehydrate ADP and P_i. If protons were the source of this energy, they might be employed, for example, to protonate the -OH of phosphoric acid to stabilize it as a leaving group.

An energy profile of this one-step reaction is shown in Fig. 28.1a. It is tempting simply to extend this profile to the enzyme catalysed reaction (Fig. 28.1b) with little alteration to the energy levels of the intermediates—and again consider a mechanism in which protons from the gradient participate directly in anhydride bond formation. The role of the enzyme here is solely to lower the *activation energy* for the reaction.

However, this ignores the capacity of an enzyme to change the energy profile of a reaction. By *differentially stabilizing ATP relative to ADP and P_i*, the ATP synthase can minimize the energy difference required for the chemical step—indeed, synthesis of ATP from $ADP + P_i$ can proceed, *on the enzyme surface*, without energy input. Clearly, energy input is required at some stage in the catalytic cycle (production of ATP free in solution still requires 55 kJ/mol) but this need not be at the bond formation step; in Fig. 28.1c energy is supplied at the stage of product (ATP) release.

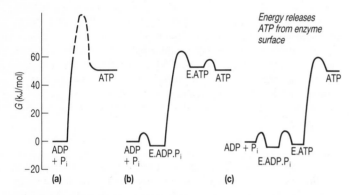

Fig. 28.1. Energy input during ATP synthesis.
  (a) Putative uncatalysed reaction (G measured relative to the starting materials, $ADP + P_i$).
  (b) Putative catalysed reaction—enzyme reduces activation energy only.
  (c) Actual mechanism—enzyme also alters relative stability of enzyme bound intermediates.

The 'binding change model' of ATP synthesis

These considerations—in particular the shift of emphasis away from 'active chemical' intermediates towards binding energy changes—has led to our present model of ATP synthase function. The conformational or binding change model runs as follows (Fig. 28.2):
1 $ADP + P_i$ bind to the catalytic site of F_1. *An equilibrium is established between $ADP + P_i$ and ATP.*
2 Because the affinity of F_1 for ATP is so much higher than that for $ADP + P_i$, *the equilibrium position on the enzyme is much more in favour of ATP than that in free solution*. Newly synthesized ATP is very tightly bound at the active site of F_1.
3 The energy (protons) supplied for ATP synthesis is employed to *decrease the affinity of the enzyme for ATP—by changing the conformation of the protein*. ATP is expelled from the active site.
4 This unstable conformation of the enzyme relaxes to the more stable, initial form.

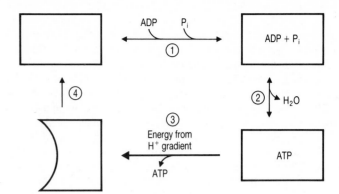

Fig. 28.2. The conformational (or 'binding change') model of ATP synthesis. The box represents one active site of the F_1 molecule.

The use of energy to *change ATP binding* affinity, by changing the *conformation* of F_1, has led to this being designated the 'binding change' or 'conformational' model for ATP synthesis. It is known that enzymes do adjust the concentrations (stabilities) of intermediates in many catalysed reactions; observations of intermediates in many kinase reactions (by nuclear magnetic resonance spectroscopy), for example, confirm that intermediates are present in roughly equal concentrations ($K_{eq} \approx 1$) on the enzyme surface, whatever the equilibrium constant in solution. Thus the above model, although initially surprising, is consistent with the energetics of reactions catalysed by other enzymes. Furthermore, the use of ion binding energy (from H^+) to counteract ATP binding energy (and hence drive ATP release from the active site) is closely analogous to mechanisms proposed for other ion pumping systems (Section 44).

Evidence supporting the 'binding change' model
The various postulates of the 'binding change' model have been independently verified.

(1) ATP binds more tightly to the ATP synthase than does ADP
This prediction derives from Fig. 28.1c, and clearly differentiates this model from that of Fig. 28.1b. ATP binding is difficult to measure directly, because it is rapidly hydrolysed by the ATP synthase. However, a non-hydrolysable analogue, AMP-PNP (Fig. 28.3), can be studied. This binds to F_1 (or F_1F_o) with $K_d \leqslant 10^{-12}$M, whereas ADP binds with $K_d \approx 10^{-5}$M. This corresponds to a difference in binding energy of $\geqslant RT \ln 10^6 \approx 35$ kJ/mol i.e. the ATP analogue (and hence ATP) is stabilized on the enzyme surface relative to $ADP + P_i$ by $\geqslant 35$ kJ/mol.

(2) ATP and ADP are in equilibrium on the surface of the ATP synthase
The equilibrium between ATP and $ADP(+P_i)$ can be probed using an *isotope exchange* method. If ATP is hydrolysed in H_2O^{18}, O^{18} is incorporated into the released P_i on bond breakage

$$ADP-O-\overset{\overset{\displaystyle O}{\|}}{\underset{\underset{\displaystyle O^-}{|}}{P}}=O + H_2O^{18} \rightarrow ADP-O^- + O^{18}-\overset{\overset{\displaystyle O}{\|}}{\underset{\underset{\displaystyle O^-}{|}}{P}}=O$$

Measurements indicate that hydrolysis of ATP by the ATP synthase yields P_i with *3 or 4 mol O^{18}/mol P_i*. This can be explained only if the terminal pyrophosphate bond is *cleaved and reformed several times* before P_i is released (see step 2, Fig. 28.2)—each cycle of cleavage

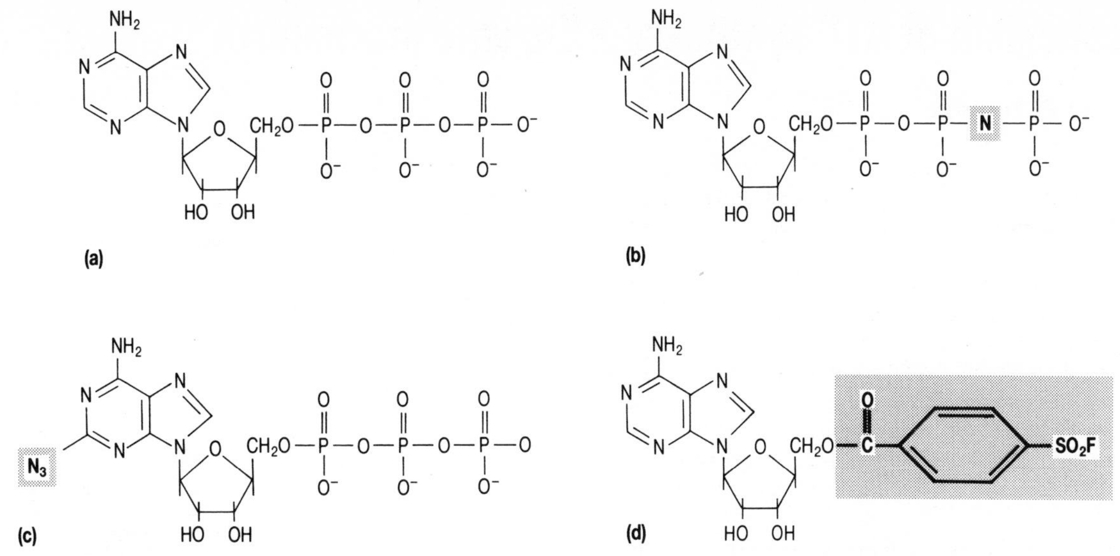

Fig. 28.3. ATP analogues used in studying F_1.
(a) ATP (see Section 7).
(b) AMP–PNP—a non-hydrolysable analogue.
(c) 2 azido ATP—a photoffinity label.
(d) Fluorosulphonyl benzoyl adenosine—an affinity label.

and bond formation incorporating more O^{18} into the terminal phosphate of ATP.

Since this occurs in the absence of any transmembrane proton gradient (and even on isolated F_1), hydrolysis and synthesis of ATP on the enzyme surface must occur without energy input; an equilibrium ($\Delta G \approx 0$) is established.

(3) The equilibrium constant for ATP cleavage on the enzyme is close to one

K_{eq} on the enzyme can be probed by independent kinetic measurements of the rate constants (k) of the forward (E.ATP → EADP.P_i) and reverse (E.ADP.P_i → E.ATP) reactions (step 2 in Fig. 28.2), using rapid reaction methods. In ox heart

$$k_{forward} = 10 \sec^{-1} \qquad k_{reverse} = 24 \sec^{-1}$$
so that $K_{eq} = 2.4$

This can be compared with a value for K_{eq} for ATP $\rightleftharpoons$ ADP + P_i in free solutions of about 10^5 M. This means the $\Delta G^{0'}$ on the enzyme surface (in addition to ΔG) is close to zero.

(4) Energy promotes ATP release from the enzyme surface

Again, AMP-PNP is used as a non-hydrolysable ATP analogue. By preincubation, [^{3}H]-labelled AMP-PNP can be loaded into the active site of the ATP synthase on submitochondrial vesicles (see (1) above). If these vesicles are given an oxidizable substrate (e.g. NADH), their membranes are 'energized'. Under these conditions, AMP-PNP is released from the active site of the enzyme (the radio-label appears in solution) in the absence of any chemical synthesis *at rates comparable with normal rates of ATP synthesis*. Thus energy linked ATP release (step 3, Fig. 28.2) is fast enough ('kinetically competent') to be a step on the pathway to ATP synthesis.

Summary

Thus, in summary, **ATP can be made on the surface of F_1 without energy input; the energy used for continuous ATP synthesis is utilized to change the conformation of F_1 to release this ATP from it. This energy is derived from proton movement. It is probable that H^+ ions cross the membrane to protonate particular acid groups on F_1, causing a conformational change in the protein.**

29 Mechanism of ATP synthesis: 2 Enzyme mechanism

Mechanism of F₁ turnover

F_1 (and in particular its β subunit) catalyses an equilibrium between ATP and ADP + P_i, with an equilibrium constant close to one. We do not know precisely which groups on the enzyme interact so much better with ATP; approaches to this problem are discussed in Section 27.

Proteins move through F_o and cause a conformational change in F_1, causing ATP to be released from its active site. How the protons traverse the membrane is discussed in Section 30.

To complete a description of ATP synthesis we need to establish:

1 How can protons crossing F_o cause a conformational change in F_1 i.e. how does F_1 'sense' the transmembrane flux of H^+?

2 How is the movement of *3* H^+ ions coupled to the synthesis of *1* molecule of ATP (see Section 23) i.e. how does F_1 'count' the protons?

3 After ATP is released from F_1, what prevents it rebinding to the high affinity site it has just left and thus inhibiting phosphorylation? The answer to none of these questions is known in detail.

How does F₁ interact with transmembrane protons?

The part of F_1 sensing the proton gradient is its γ subunit; damage to this subunit may lead to either a decreased or increased flux of protons, suggesting this may represent a proton gate (Section 27). Furthermore, the existence of a slender stalk (surrounded by water) connecting F_1 to F_o suggests that protons are not conducted through the membrane directly to the active site of F_1, but that their passage through the channel is transmitted, via an (obligatory) conformational change in the a or b subunit of F_o (Section 30), to the γ subunit of F_1. Thus protons crossing F_o protonate an (internal) F_o peptide, changing its conformation; this interacts with F_1 peptides (ultimately the β subunit), causing a change in conformation at the active site and hence ATP release.

Unidirectionality of ATP release

This phenomenon may be explained by the unusual enzymatic mechanism of F_1—the alternating site mechanism.

F_1 contains 3 catalytic sites, one on each of its 3 β subunits. In a catalytic cycle, each site must adopt 3 conformations, viz.

1 T(ight): ATP and ADP + P_i are in equilibrium at the active site. ATP is bound tightly ($K_d < 10^{-12}$ M) and the catalytic groups are in place. Both ATP and ADP exchange only very slowly with the bulk solution.

2 O(pen): ATP is expelled from the active site (K_d rises to $> 10^{-4}$ M), due to a conformational change. The catalytic groups are displaced, and this state is catalytically inactive.

3 L(oose): ATP and ADP + P_i bind equally well to the enzyme (there is no differential binding), and can exchange rapidly with solution nucleotides. Again, catalysis is absent, and so no equilibrium is established.

A catalytic cycle in ATP synthesis, depicted in Fig. 29.1, requires each conformation to be taken up sequentially. The behaviour of one catalytic unit, which passes through the conformations $T \rightarrow O \rightarrow L \rightarrow T \ldots$ is shown shaded in Fig. 29.1.

In F_1, the 3 catalytic sites do not pass through the three conformations randomly, but are *linked together 120° out of phase*. At any instant, the enzyme will possess 3 catalytic sites each of which is in a different conformation; a moment later, the enzyme will take up the same (mixed) conformation overall but each individual site will have switched to the next stage in its catalytic cycle (Fig. 29.1).

In this mechanism, the tendency of released ATP to rebind to F_1 is eliminated by permitting nucleotides to exchange with the solution only in conformation L, which has no preference for ATP; a tight

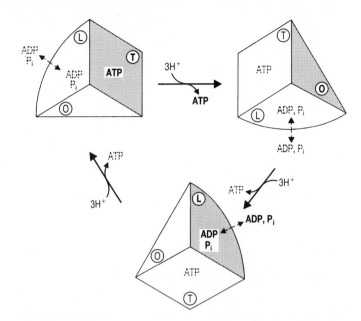

Fig. 29.1. Alternating site mechanism for ATP synthase. Sequential changes in one catalytic unit shown shaded. Each step represents a cycle of H^+ binding/ATP release as shown in Fig. 28.1, with each catalytic unit at a different stage in the cycle (120° out of phase). Note that all three conformations of F_1 are identical, but rotated through 120° anticlockwise.

ATP binding site (T) can be formed only *after* an active site is occupied by ADP and P_i. Similarly, the alternation of *three* conformations may provide a link to the counting of *three* protons.

Evidence supporting an alternating site mechanism

Negative cooperativity in ATP binding

Despite there being 3 catalytic sites on F_1, which are in identical positions on identical β subunits, subsequent molecules of ATP (or its analogue, AMP-PNP) bind with decreasing affinity. K_d values for AMP-PNP range from $< 10^{-12}$ M (T) to 10^{-6} M (L) and 10^{-4} M (O conformation). Similarly, 3 K_m values (showing 3 decreasing affinities) for ATP hydrolysis can be observed on isolated F_1.

Positive cooperativity in enzyme turnover

At very low ATP concentrations ($< 10^{-6}$ M), only the T site is occupied by nucleotide. Under these conditions, catalytic turnover (ATP hydrolysis) is very slow because neither ATP nor ADP can be easily released from the enzyme (see above). The catalytic cycle requires the L site to be filled before nucleotide can be exchanged with the solution. At higher concentrations of ATP, the O site fills and turnover can proceed rapidly. Thus 3 V_{max} values (showing 3 increasing turnover rates) for ATP hydrolysis can be observed on F_1.

Conformational changes in F₁

One suggestion for linking a proton gate with the alternating site model is given in Fig. 29.2.

1 The single γ subunit is in contact with one of the 3β subunits. This subunit is in the O form.

2 Three protons entering F_o (c subunit) induce a *conformational change in the b subunit* of F_o, which is in contact with $F_1(\gamma)$.

3 A change in conformation in the γ subunit ensues (indicated in Fig. 29.2a), *driving it off one β subunit onto the next*. The newly

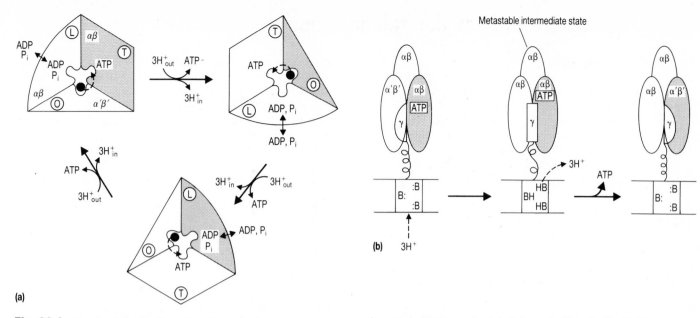

(a)

Metastable intermediate state

(b)   3H⁺

Fig. 29.2. F₁ subunits in the alternating site mechanism.

(a) Top view, showing possible role of the γ subunit in generating asymmetry. The catalytic unit is designated as an αβ subunit pair; the pair

associated with the γ subunit is designated α'β' (as in Fig. 26.1d).

(b) Side view, showing one conformational transition. The shaded subunit changes from T → O form.

contacted β is driven into the O conformation, and the other β subunits change accordingly.

4 The movement of the γ (and hence the b subunit) *releases the 3 protons from F_o on the F_1 side of the membrane*. The three proton binding sites are now free, and the cycle can repeat.

This model is speculative, based on the asymmetric appearance of F₁ in high-resolution electron micrographs (Section 27) and, to

some extent, analogies with the mechanism of the H⁺ driven flagella motor (Section 39). (Note that the F₁ rotates relative to the γ subunit—or vice versa.) The energy released when the γ subunit of F₁ binds to the new β subunit is used to drive ATP of the β subunit. Note that the transmitted protons need not interact directly with F₁ to induce its conformational change.

30 F_o—structure and function relationships

Polypeptide composition of F_o

Like F_1, F_o is a complex of several polypeptides, designated (in *E. coli*) a, b and c. These are all transmembrane polypeptides. Again, there are multiple copies—1 a, 2 b and a large number (probably 10–12) c subunits make up one F_o moiety. Its 'active site' (the proton carrying site or 'protonophore') is located on the c subunit. The *energy transfer inhibitor*, DCCD (see Section 26) prevents H^+ flow; its site of covalent binding is an acidic residue on this, the smallest subunit of the entire F_1F_o complex (mol. wt. $\approx 8\,kDa$).

Structure of subunit c

Subunit c is a small, very hydrophobic peptide. It is soluble in solvents typically used for lipids (chloroform/methanol) and is sometimes called a *proteolipid*. It folds to traverse the membrane twice, the *two transmembrane helices* being joined by a short loop (rich in gly, pro and charged residues), protruding out of the membrane on the side of F_1 (Fig. 30.1).

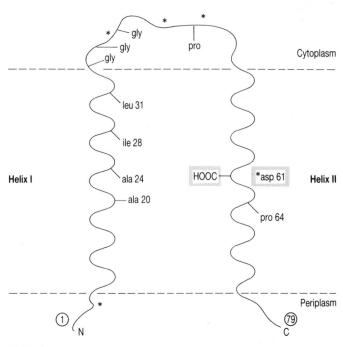

Fig. 30.1. Subunit c of F_o, showing residues named in text and charged residues (*).

The **protonophoric group** is a carboxylic acid buried in the membrane phase, on the C-terminal helix (helix II). In *E. coli*, this residue is **aspartic acid 61**, although in most other organisms it is glutamic acid. This is the residue which reacts with DCCD, blocking H^+ translocation. Mutants where this group is replaced (e.g. by asparagine or glycine) cannot translocate protons.

The two helices pack closely together in a *hairpin* structure. Evidence for this arrangement comes from the demonstration that mutations in helix I affect the function of the protonophore on helix II. First, altering the size of residues 28 (ile → val) or 24 (ala → leu) on helix I prevents DCCD reacting with asp 61 ('DCCD-resistant mutants'). Secondly, mutations in helix I can complement (restore activity to) deleterious mutations in helix II (e.g. ala 20 → pro restores activity to the pro 64 →leu mutant). Indeed, H^+ transfer can proceed if the carboxyl group is moved across to helix I (as in the double mutant asp 61 → gly, ala 24 → asp) suggesting a virtual equivalence of the two arms in space.

The next level of structure involves the assembly of several c subunits into F_o which contains 10–12 of them. They appear to pack in a specific arrangement, since only one face of each helix is exposed to the membrane phase (as shown by labelling with a lipid-soluble reagent). The other face (inaccessible to label) is presumably in contact with other c subunits, the a and b subunits of F_o, and/or a water filled pore. However, little is known of how the several hairpins interact; no amino acid residues likely to interact are obvious by inspecting the sequence. Comparing sequences between species, we find that (i) hydrophobic residues remain hydrophobic, but there is no strict conservation of individual amino acids and (ii) several glycine residues, especially in helix I, are conserved, possibly indicating some limitation of space in the packed assembly.

Structure of subunit b

There are 2 copies of subunit b per F_o proton channel, and it is this subunit that participates in the F_1-binding stalk. The b subunit has a single transmembrane helix at its N-terminal end. The remaining 80 per cent of the molecule is extramembranous, and probably comprises 2 α helical sections which contain many charged groups (Fig. 30.2a). The length of these helical sections may be as long as 10 nm, certainly long enough to traverse the stalk region and contact the bulk of F_1 (Fig. 26.1c).

Structure of subunit a

Subunit a lies largely within the membrane, having six or seven transmembrane helices (Fig. 30.2b). One of these helices (helix V)

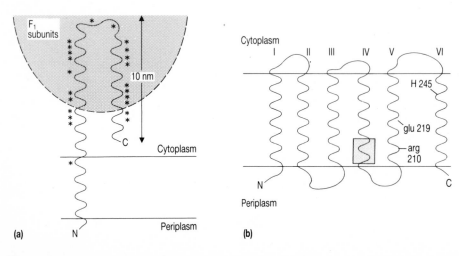

Fig. 30.2. (a) Subunit b of F_o, showing distributions of charged residues (*).
(b) Structure of subunit a, showing residues critical for H^+ transfer and region where oligomycin binds (shaded area). H = histidine.

has a polar side (it is *amphipathic*), indicating that it makes polar interactions within the membrane. It is possible that this region of subunit a interacts with asp 61 of subunit c, adding to the structure of a proton 'pore' in the membrane (Fig. 30.3a). The sequences of helices V and VI are the most highly conserved between species, and mutation indicates that arginine 210, glutamate 219 (helix V) and histidine 245 (helix VI) are required for proton transfer. A 'proton relay' system has been suggested (Fig. 30.3a).

(Studies on mitochondrial mutants implicate helix IV of subunit a as the site of interaction of F_o with the proton translocation inhibitor, oligomycin (Section 26). This site is shaded in Fig. 30.2b. This compound does not act on the *E. coli* enzyme.)

Subunit a also interacts with b, allowing the effects of proton movement in F_o to be transmitted (via b) to F_1. Again, it is possible that helix VI is involved, since mutation at proline 240 affects a–b interaction.

Mechanism of proton translocation

Protons cannot cross the lipid bilayer of a membrane unaided. There are two possibilities—they cross by 'hopping' from base to base within the F_o proteins ('proton relay') (Fig. 30.3a), or they pass (probably as H_3O^+) through an aqueous pore formed, and given its specificity, by the surrounding protein (Fig. 30.3b). This latter is more analogous to other ion pumping mechanisms (Section 43). It is not clear which (if either) mechanism operates in F_o, and the details, therefore, are highly speculative.

Sequence of the F_1F_o subunits—from gene to protein

The determination of the protein sequences for the subunits of F_1 and F_o was a landmark in the study of the ATP synthase and its mechanism. The methodological approach is given below.

1 Mutants in ATP synthesis (*un*coupled or *atp* mutants) were obtained in *E. coli* by selecting bacteria able to grow on fermentable substrates (e.g. glucose) but not substrates requiring oxidation (succinate, glycerol).

2 Genetic mapping indicated 8 structural genes located in the single operation in *E. coli* (Section 26).

3 A plasmid containing this operon was used to amplify the number of gene copies and provide material for DNA sequencing.

4 Reading frames were identified and translated into protein sequence.

5 The sequence was checked against partial sequences obtained from direct protein sequencing (e.g. of the region around the site of labelling by an affinity label).

6 Sequences were inspected for hydrophobicity, and transmembrane segments predicted (see above). Database searches revealed homologies with other proteins (e.g. ATP binding regions in the β subunit, Section 27).

7 The DNA sequence provided genetic probes useful for isolating

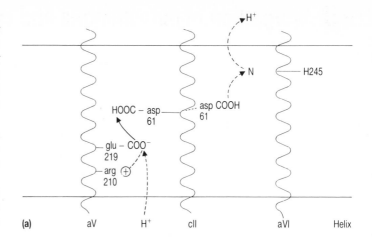

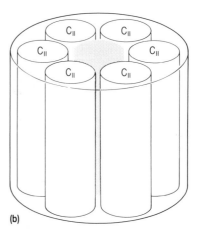

Fig. 30.3 (a) Possible H^+ relay pathway through F_0 a and c subunits. (b) Possible construction of a pore from F_o–c subunits. Only helix II is shown; asp-61 points into the (water filled) channel.

homologous genes from other sources (e.g. man, cow, pea etc.), where mutants are not available, and the genes are not contiguous.

8 Comparative sequencing of homologous proteins yielded information about amino acid residues essential for function.

9 Analysis of functional defects in random, and site directed, mutants enabled mechanistic models to be tested.

10 (X-ray analysis reveals the three-dimensional structure of the protein—still to come for this protein!) *

* The X-ray structure of bovine heart F_1 has recently been solved, providing insights into the catalytic mechanism (Abrahams *et al.*, 1994).

31 Integration of mitochondria and cytoplasm

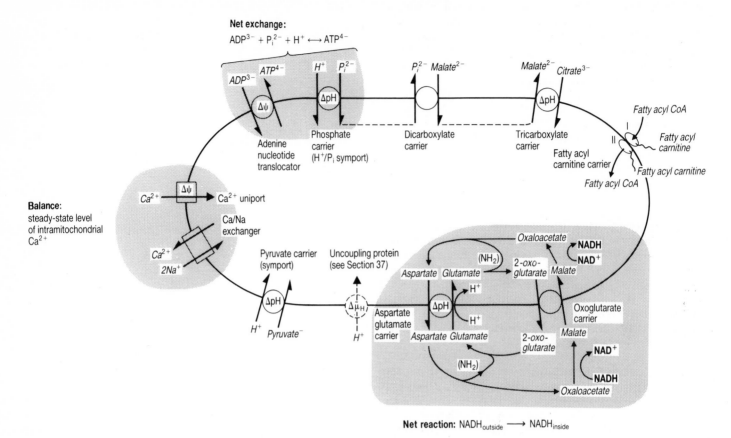

Fig. 31.1. System of mitochondrial ion carriers, showing inter-relating systems.

Compartmentation

In bacteria, ATP is made and used in the cytoplasm. In eukaryotes, however, ATP synthesis occurs in mitochondria, while it is used (largely) outside. Thus *ATP must be transported out* of the mitochondrion as it is made; clearly *ADP, P_i and oxidizable substrates must be transported inside*. These polar molecules cannot cross the lipid membrane unaided; thus specific transport proteins (**translocators**) are required. Further, some molecules are moved against a concentration gradient; thus *energy (derived from $\Delta\tilde{\mu}_{H^+}$) is required*.

The separation of two compartments within a cell, although leading to the above complications, has its useful consequences. The use of energy to drive ion movements means that different concentrations of metabolites can be maintained within different cell compartments. In particular, the [ATP]/[ADP] ratio is maintained at about 10^2 in the cytoplasm (where ATP is used), but at about only 5 in mitochondria where it is made. Similarly, the [NAD$^+$]/[NADH] ratio in the cytoplasm is about 10^3 (where NAD$^+$ is used to oxidize substrates), and 10 in the mitochondria (where NADH is a reductant). This makes it possible for reactions such as oxidations and reductions which require conflicting conditions to occur in the same cell.

Phosphate-proton symporter

The most active transporter in the mitochondrial membrane is the phosphate transporter. This imports P_i into the mitochondrion. Movement of P_i inwards is accompanied by H$^+$, one proton moving inwards for each phosphate ion transported (see Section 21). The P_i transporter is thus a P_i/H$^+$ **symport**. Since H$^+$ moves into mito-

chondria down its electrochemical gradient, the *energy required for P_i movement is supplied directly by the proton gradient* (ΔpH; Fig. 31.1).

Adenine nucleotide translocator (ATP/ADP antiporter, AdN translocase)

The AdN translocase imports ADP into the mitochondrion and exports ATP. As it carries a different molecule in each direction—in a sequential manner—it constitutes an **antiport** system. It can select cytoplasmic ADP from ATP (whose concentration is much higher) by virtue of its much lower K_m^{ADP} on the cytoplasmic face of the membrane. On the inside, selectivity is lower ($K_m^{ADP} \approx K_m^{ATP}$) and ATP is exported, as a result of its higher concentration.

The net result of ATP–ADP exchange is the movement of ATP from a low concentration inside the mitochondrion to a higher concentration outside. Energy is thus required. Since *ATP has a higher negative charge than ADP* (ATP^{4-}, ADP^{3-} at pH $\approx$ 8), this energy can be supplied from the electrical part of $\Delta\tilde{\mu}_{H^+}$, the *membrane potential* ($\Delta\psi$). This is negative inside, favouring the import of the less negative species, ADP (Fig. 31.1).

Since the charge on P_i is more than unity (P_i^{2-} at pH 8.0), P_i moves *against* the electric field. However, together the adenine nucleotide and P_i transporter form a *charge balanced system* with a net stoichiometry (Fig. 31.1) of

$$ADP + P_i + H^+ \text{ inwards} : ATP \text{ outwards}$$
$$3^- + 2^- + 1^+ = 4^- \text{ inwards} : 4^- \text{ outwards}$$

This has the consequence that, for intact mitochondria to synthesize cytoplasmic ATP, one proton is needed in excess of the 3H$^+$/ATP required by the ATP synthase. *The apparent H⁺/ATP ratio = 4, 3 protons being used by the ATP synthase and 1 by the export system* (see Section 21).

Other transporters

The above transporters utilize the different components of $\Delta\tilde{\mu}_{H^+}$ directly as sources of energy. This is true of two other carriers, the *pyruvate-proton symport* (which uses ΔpH) mediating pyruvate entry into mitochondria, and the *Ca^{2+} uniport* (see Section 33), which uses $\Delta\psi$, mediating Ca^{2+} entry (Fig. 31.1). As mitochondria do not consume cations, export systems are also present to maintain a steady state level of cations. These are, typically, antiport systems; in heart, Ca^{2+} export is driven by $2\,Na^+$ import (Fig. 31.1), but H^+ may be used in other tissues.

Other carriers use $\Delta\tilde{\mu}_{H^+}$ less directly. Malate and glutamate bring reducing equivalents *into* the mitochondria, generating NADH in the tricarboxylic acid cycle. Aspartate, citrate and 2-oxoglutarate carry carbon into the cytoplasm for gluconeogenesis and fatty acid biosynthesis. These are carried on a series of **anion carriers**, using *other ion gradients* as their energy source. One can envisage a hierarchy of antiporters; malate is accumulated in exchange for P_i, citrate and 2-oxoglutarate in exchange for malate, and glutamate for aspartate (Fig. 31.1). A combination of the malate/α-oxoglutarate and glutamate/aspartate carriers, in addition, participates in a **redox shuttle**, the net result of which is to move cytoplasmic NADH (generated in glycolysis) into the mitochondrion for oxidation (Fig. 31.1). This is an uphill process ($NADH/NAD^+$ is *lower* in the cytoplasm than in the mitochondria) and energy is provided by movement of H^+ downhill along with glutamate.

Two other transporters are shown in Fig. 31.1. Fatty acids, unlike the simple organic acids above, are insoluble in water and exist in the cytoplasm and inside the mitochondrion as fatty acyl CoAs. For transport they are transferred onto carnitine via carnitine acyl transferase I, transported across the membrane via the **carnitine transporter**, and then transferred back onto CoA at the inner membrane surface. Although this process seems rather different to those described above, it represents an antiporter system (acylcarnitine in, carnitine out) and the protein involved is similar in structure to those transporting the small organic acids (see below). The other transporter depicted is the 'uncoupling protein'. This is found in brown adipose tissue, and allows H^+ to move down an electrochemical gradient, generating heat. It is thus a **H^+ uniport**. Its physiological role is discussed further in Section 37.

Structure of transporters

Despite their *varied substrates*, all the anion transporters in Fig. 31.1 (including the P_i and pyruvate symporters), the fatty acid transporter and the uncoupling protein have *related structures*. They belong to a *family* of proteins. (The Ca^{2+} transporters are unrelated.) This remarkable finding is based on comparison of their protein sequences, which show that they all have similar (homologous) organization.

Each carrier contains a single type of polypeptide chain, close to 300 amino acids long. Within this chain, we can identify a base unit of 100 amino acids, repeated three times. Each base unit has a transmembrane helix at each end, joined by a loop of 45 amino acids; thus the overall polypeptide crosses the membrane six times.

A possible structure is shown in Fig. 31.2. The N and C termini of the protein lie on the cytoplasmic side, as do the short loops joining each base unit; the large (45 amino acid) loops (within each base unit) face the matrix. Helices 2, 4 and 6 have a hydrophilic side (*amphipathic helices*), possibly lining an aqueous pore. It is believed that two such polypeptides participate in one translocator, forming a dimer (Fig. 31.2c); the transported molecule could pass through the aqueous pore made from the six amphipathic helices in a dimer.

Onto this basic structure must be superimposed the specificity of each individual carrier. It is surprising that the transport of compounds so different as H^+ (in the uncoupling protein), ATP/ADP (adenine nucleotide translocator) and fatty acyl carnitine (carnitine carrier) can be accomplished with a single structural framework. Presumably, the key to specificity lies in the large extramembrane loops on the matrix side, and the three-dimensional structure they adopt. However, this is not yet known.

The existence of a repeating, 100 amino acid, base unit in the structure of these translocators indicates that it may have arisen from gene duplication and fusion of an original 100 amino acid polypeptide. Sequence homology between the three units is limited, but does support this view; there is, for example, a conserved proline at the end of helix I, and glycine at the start of helix II in each base unit (Fig. 31.2a) in all translocators studied.

This family of transporters, however, is strictly a *mitochondrial family*. It bears no structural relationship to bacterial transport systems (Section 38) or even to the chloroplast P_i transporter (Section 36), indicating that it evolved after divergence of mitochondria from free living precursor forms (Section 45).

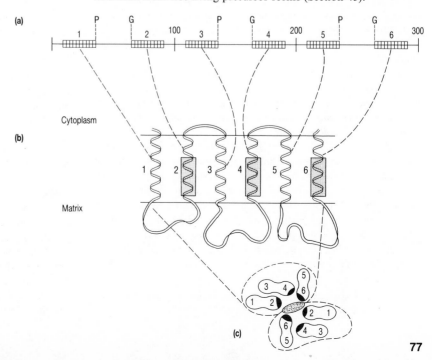

Fig. 31.2. Structure of mitochondrial transporter family.

(a) Tripartite structure of carrier showing 100 amino acid repeat. Transmembrane regions are boxed. P = proline, G = glycine; conserved amino acids.

(b) Orientation of polypeptide in the membrane. Amphiphilic helices are shaded.

(c) Possible association of two carrier molecules to produce a hydrophilic pore. View from matrix side.

32 Control of ATP synthesis: 1 Thermodynamic aspects

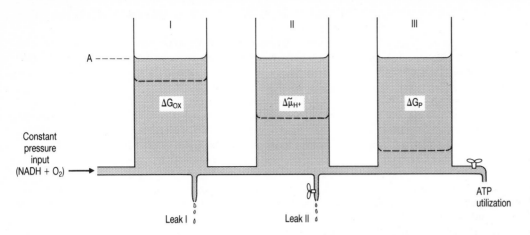

Fig. 32.1. Hydrodynamic analogy for control of respiration by ATP utilization (near-equilibrium model). Leak I represents energy lost as heat in oxidation of cyt c by O_2 (non-equilibrium step). Leak II represents energy lost by H^+ leakage across the membrane.

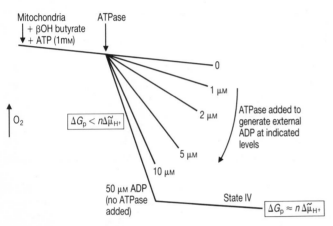

Fig. 32.2. Dependence of mitochondrial respiration rate on external [ADP].

Respiratory control

Mitochondria use oxygen at a rate dependent on the availability of ADP and P_i (Section 22). In the presence of an oxidizable substrate alone, ΔG_{ox}, the free energy change on oxidation, is highly negative, oxidation is nonetheless very slow (state IV respiration). When ADP and P_i are added, oxygen uptake speeds up—and this increased rate (state III respiration) persists until the added ADP has been converted to ATP, when the slow state IV rate returns (Fig. 22.3). The influence of ADP and P_i on oxidation rate is known as **respiratory control**.

This remarkable observation demonstrates that *the respiratory chain senses changes in ATP synthase activity*—the two processes appear to be **coupled** together. This occurs because energy yielding oxidations and energy utilizing phosphorylations *communicate through a common intermediate*, $\Delta\tilde{\mu}_{H^+}$.

Oxidation rates thus vary as follows. Oxidation continues until the (redox) proton pumps are halted by the back pressure of protons (state IV). That is to say, the proton pumps, *come to equilibrium with* $\Delta\tilde{\mu}_{H^+}$ and (by the definition of equilibrium), net oxidation ceases.*
When ADP and P_i are introduced, protons are used to make ATP, $\Delta\tilde{\mu}_{H^+}$ decreases in magnitude, and oxidation—no longer pushing against such a high back pressure—can proceed (state III).

Ultimately, of course, so much ATP is made that ΔG_P, the free energy for ATP hydrolysis, just balances $\Delta\tilde{\mu}_{H^+}$; no more ATP

---

*A low, rather than zero, residual rate is observed, which reflects the leakage of H^+ across the membrane.

synthesis is possible (equilibrium) and $\Delta\tilde{\mu}_{H^+}$ again limits oxidation (state IV).

This can be illustrated by the hydrodynamic analogy shown in Fig. 32.1. The constant pressure input can support a level A in reservoir I; in the absence of outflow, liquid in reservoirs II and III will reach the same level and then input will cease, due to back pressure. If, however, liquid is drawn off from reservoir III, levels in I and II will fall (dotted lines) and input will occur.

The idea that $\Delta\tilde{\mu}_{H^+}$ is directly responsible for the limit on oxidation rate can be confirmed; uncouplers, which decrease $\Delta\tilde{\mu}_{H^+}$ independently of ADP, also speed up respiration (Section 34). This is equivalent to increasing leak II in Fig. 32.1.

Since changes in oxidation rate are brought about by limited displacement of $\Delta\tilde{\mu}_{H^+}$ from an equilibrium with the redox pumps, this model is termed the **near-equilibrium model** for respiratory control.

Equilibrium and non-equilibrium reactions in the respiratory chain

This model is somewhat oversimplified when applied to real systems. As indicated by (i) measurements of redox potential and (ii) reversibility of electron flow, the redox span from NADH to cyt c may well approach equilibrium with $\Delta\tilde{\mu}_{H^+}$—in accordance with this model. However, as electrons pass from cyt c to O_2, the free energy lost, (ΔG), is significantly larger than the energy conserved at this step $(\Delta\tilde{\mu}_{H^+} \times H^+$ pumped).

This has the advantage that the flow of electrons to oxygen (from NADH etc.) is always downhill—maintaining a flux through the system. However, as regards our model for control, only the span from NADH to cyt c can reach equilibrium and be subject to near equilibrium control. Cyt aa_3 must operate relatively slowly so as to maintain this equilibrium; it cannot reach equilibrium itself. This slow rate might be due to the low intrinsic activity of the enzyme and/or a low concentration of its substrate, reduced cyt c. It is represented by leak I in Fig. 32.1.

Rates of respiration and ATP synthesis

In vivo, ATP synthesis is rarely either fully on or fully off. States III and IV represent, in fact, two extreme rates of oxidation between which a graded response occurs. The actual rate observed depends on ADP *concentration*; this can be varied artificially (*in vitro*) by adding varying amounts of an ATPase to a phosphorylating mitochondrial suspension. As ATPase activity is increased, the steady state level of ADP rises, and the oxidation rate increases (Fig. 32.2). Presumably, in the cell, a similar situation applies; as ATP use

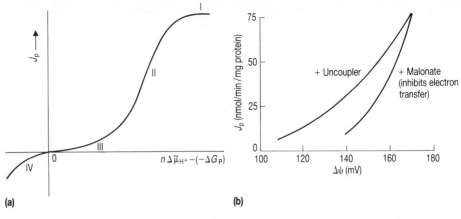

Fig. 32.3. (a) Theoretical relationship between ATP synthesis rate (J_P) and driving force [$n\Delta\tilde{\mu}_{H^+}-(-\Delta G_P)$].

(b) Measurements of ATP synthesis rate as a function of $\Delta\psi$ (ΔG_P and ΔpH kept low and constant), using different reagents to vary $\Delta\psi$.

increases, for example, ATP is resynthesized faster in response to increased steady state levels of [ADP]. This method of regulating ATP synthesis seems to apply in **skeletal muscle** recovering after a burst of intense, anaerobic activity. Nmr measurements here indicate that oxidation rates increase with [ADP] in the cytoplasm.

The response of oxidation rates to ADP concentrations is not a linear one; at its simplest, we might expect the ATP synthase to saturate at high levels of ADP and so, above a certain rate (state III respiration), no further increase would be observed (Fig. 32.2). However, because the mitochondrion is a complex assembly of enzymes, simple enzyme kinetics cannot describe their behaviour (K_m^{ADP} varies with $\Delta\tilde{\mu}_{H^+}$, for example), and other approaches are required for a quantitative description. These approaches may be classified, roughly, as thermodynamic (below) or kinetic (see Section 33). The thermodynamic treatment is based, implicitly, on the near-equilibrium model for control, as defined above.

Quantitative aspects of the near-equilibrium model

The observations above can be restated, in thermodynamic terms, as follows. ATP synthesis proceeds at a rate ('flux') J_P, which depends on the electrochemical gradient ($\Delta\tilde{\mu}_{H^+}$) and the energy required to make ATP ($-\Delta G_P$ where ΔG_P is the free energy of ATP *hydrolysis*; Section 23). If n protons are used to make ATP:

1 ATP will be made (J_P +ve) when $n\Delta\tilde{\mu}_{H^+} > -\Delta G_P$.
2 ATP will be hydrolysed (J_P –ve) when $n\Delta\tilde{\mu}_{H^+} < -\Delta G_P$.
3 The system will be at equilibrium ($J_P = 0$) when $n\Delta\tilde{\mu}_{H^+} = -\Delta G_P$. On Fig. 32.1, these are indicated by: (1) level in reservoir II being higher than reservoir III (dotted lines); (2) reservoir III higher than reservoir II (not shown); and (3) reservoir II equal to reservoir III (solid lines).

The rate of ATP synthesis, J_P is thus related to the **net** thermodynamic driving force for synthesis [$n\Delta_{H^+}-(-\Delta G_P)$]. Irreversible thermodynamics suggests that a fixed mathematical relationship holds, viz.

$$J_P = f[n\Delta\tilde{\mu}_{H^+}-(-\Delta G_P)]$$

where f represents some mathematical function. It does not, unfortunately, predict the form of this relationship.*

Measurements of such *flow–force relationships* in mitochondria yield complex curves of the type shown in Fig. 32.3a. The relationship is continuous (a small change in $\Delta\tilde{\mu}_{H^+}$ gives a small change in J_P), with four distinguishable regions from right to left:
(i) Plateau: further increases in $\Delta\tilde{\mu}_{H^+}$ give no further change in rate because the *maximum turnover rate* of the ATP synthase has been reached (kinetic limitation).

(ii) A steep rise in J_p with $\Delta\tilde{\mu}_{H^+}$. Over this range, a simple relationship $J_p = k [n\Delta\tilde{\mu}_{H^+}-(-\Delta G_P)]$ (flow proportional to driving force) is approximately true. (Thermodynamic limitation.)
(iii) A flatter region near the ordinate. This reflects the lack of H^+ (substrate) required by the ATP synthase. Three H^+ are required per ATP—hence the very sharp dependence of the curve on H^+ (probably $[H^+]^3$). (Kinetic limitation.)
(iv) As $n\Delta\tilde{\mu}_{H^+}$ falls below ΔG_P, J_P becomes negative and ATP is hydrolysed.

Implications of flow–force relationships

The demonstration, in *isolated mitochondria*, of a flow force relationship of the type shown in Fig. 32.3a is consistent with the ideas that:
1 $\Delta\tilde{\mu}_{H^+}$ drives ATP synthesis (since it is predicted from relationships between $\Delta\tilde{\mu}_{H^+}$ and ΔG_p), and
2 respiratory control can be explained on the near-equilibrium model (above).

It does not indicate, however, that respiration rates *in vivo* are under near-equilibrium control. First, it has not yet been proven that [$n\Delta\tilde{\mu}_{H^+}-(-\Delta G_P)$] *in vivo* lies in the range where J_P is sensitive to the thermodynamic driving force (region (ii) in Fig. 32.3a). Secondly, factors such as substrate (e.g. NADH) supply and enzyme activities, which are omitted in this treatment, must be taken into account. These are dealt with explicitly in the next section.

Flow–force relationships and localized protons

One aspect of flow–force relationships has been the subject of considerable dispute. The above treatment implies that *only a single relationship exists between J_P and driving force [$n\Delta\tilde{\mu}_{H^+}-(-\Delta G_P)$]* i.e. that however $\Delta\tilde{\mu}_{H^+}$ is varied (e.g. by adding ADP, uncouplers, or inhibitors of electron flow), the curve of Fig. 32.3a is unchanged. In fact, the curve does change if the mitochondria are manipulated differently (Fig. 32.3b).

This has been taken by some workers to indicate that the energy stored in the proton gradient $\Delta\tilde{\mu}_{H^+}$, cannot *directly* drive phosphorylation: that the actual driving force is different, possibly a *pool of localized protons within the membrane*. In this case, J_P and $\Delta\tilde{\mu}_{H^+}$ are only adventitiously related, under some circumstances.

A less drastic explanation for this anomaly lies, again, in consideration of enzyme activities. J_P *must be related, not only to driving force, but also the activity of the ATP synthase*. If this enzyme were switched off, for example, no matter how high was $\Delta\tilde{\mu}_{H^+}$, ATP synthesis rates would be zero! Variations in flow–force relationships probably reflect direct control of enzyme activity—e.g. modulation of the ATP synthase (see Section 33). This is analogous to varying the bore in the tube between reservoirs II and III in Fig. 32.1; the flow between these reservoirs will depend not only on the pressure difference between them but also on the bore of the connecting tube.

---

* For physical processes, such as diffusion, $J = k(\Delta\tilde{\mu})$ where k is a numerical constant i.e. the relationship is a linear one—over a wide range of $\Delta\tilde{\mu}$. For chemical changes, this is rarely true, since $\Delta G^{\ddagger}$ (activation energy), not ΔG, is the major factor in determining rates (see Section 3).

33 Control of ATP synthesis: 2 Kinetic aspects

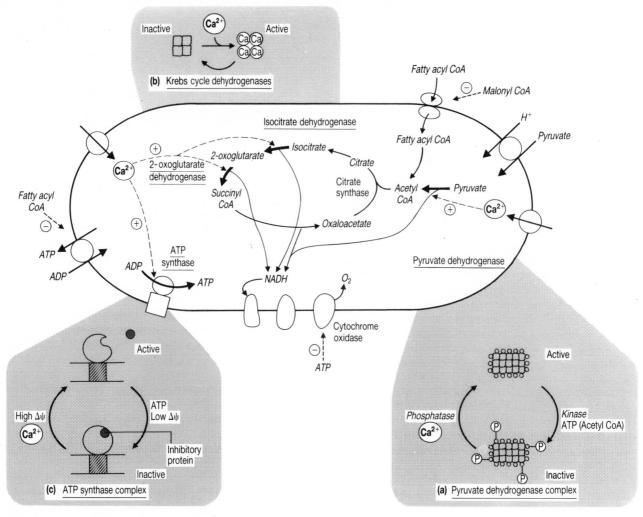

Fig. 33.1 Regulatory enzymes in oxidative phosphorylation. Those shown underlined are regulatory in heart. Others named may be regulatory in liver.

Equilibrium and non-equilibrium 'control'

Within the cell, ATP synthesis rates must always balance ATP utilization. In principle, the near-equilibrium mechanism described above could bring about this balance; as cytoplasmic ADP levels rise, ATP is made faster and, via interactions with $\Delta\tilde{\mu}_{H^+}$, respiration is speeded up. No special 'regulatory' properties of the enzymes are required; the ATP synthase simply responds to the concentration of its substrate, ADP, and the respiratory chain to its 'product', $\Delta\tilde{\mu}_{H^+}$.

This mode of regulation probably occurs *in vivo* to some extent (see Section 32). However, it cannot represent the whole story.

1 Only ADP can stimulate respiration on this model. Thus only a retrospective response is possible (respiration would not increase until *after* ATP has been used). However, hormones can stimulate respiration without any increased utilization of ATP.

2 Nmr studies in perfused heart indicate that respiration rates can vary over a fivefold range with no increase in cellular [ADP]. Thus ADP cannot be the sole determinant of respiration rate.

3 Speeding up oxidation would very rapidly deplete the mitochondrion of NADH, and the process would simply slow down again. Thus NADH *production* must be increased in line with its *utilization*.

It is likely, therefore, that oxidative phosphorylation is largely regulated by agents other than ADP. By analogy with other metabolic pathways, we would expect regulation of a few 'control' enzymes whose activity is normally low, i.e. they form 'bottlenecks' in the pathway—and whose regulators are not simply substrates or products of their own reactions.

Such **regulatory enzymes** *cannot be operating at equilibrium*; if they were, substrate and product levels (and hence substrate → product flux) could not be controlled independently. Thus regulation of this type is designated '*non-equilibrium control*', to contrast with near-equilibrium control as described in Section 32.

Controlling metabolic pathways

To apply these ideas, from metabolic control theory, to mitochondria, we must

1 identify the 'controlling' or 'rate limiting' enzymes,

2 identify possible physiological regulators of these enzymes, and their effects,

3 investigate the concentrations of these regulators in the cell, under different metabolic states,

4 develop a quantitative model to assess which enzymes and regulators contribute significantly to control *in vivo*.

Because mitochondria are closed structures, with their own internal environment, not all these studies have been completed. Nonetheless, a reasonable model for regulation of respiration and ATP synthesis can be proposed.

Control of mitochondrial respiration in heart cells

Identification of non-equilibrium reactions, and/or enzymes with a non-zero control strength, has suggested a large number of potential control sites in mitochondria. These are shown in Fig. 33.1, together with putative regulators (where known). In heart cells, substrate carriers do not seem to limit respiration, and the primary sites of regulation are the α-oxoacid dehydrogenases (underlined). All these three enzymes are activated by Ca^{2+} ions, although the mechanism varies between them.

Pyruvate dehydrogenase does not bind Ca^{2+} directly; Ca^{2+} activates an intrinsic phosphatase which dephosphorylates the dehydrogenase, *activating this* in turn. If Ca^{2+} is removed, phosphatase activity falls and the dehydrogenase is inactivated by an intrinsic kinase ($+$ATP) (Fig. 33.1a).

Isocitrate dehydrogenase and **2-oxoglutarate dehydrogenase** are activated directly (allosterically) by Ca^{2+} binding (Fig. 33.1b).

Thus, *if the heart is made to beat faster*, cytoplasmic $[Ca^{2+}]$ rises due to increased release from the sarcoplasmic reticulum. Ca^{2+} moves across the mitochondrial membrane via the Ca^{2+} uniport (see Section 31), so intramitochondrial $[Ca^{2+}]$ rises and the dehydrogenases are activated. As a result [NADH] rises inside the mitochondrion.

NADH is a substrate for the respiratory chain, and it is believed that, as NADH dehydrogenase is unsaturated, a rise in [NADH] will bring about an increase in oxidation rate. By this mechanism, contraction leads to increased respiration. The same mechanism holds for *hormones, such as adrenaline, which stimulate respiration* (in advance of actual need) by opening a plasma membrane Ca^{2+} channel and allowing Ca^{2+} into cells.

A secondary site of regulation in heart is the ATP synthase. As respiration is speeded up the ATP synthase must also increase its activity to keep pace. Increases in levels of its substrates ADP and $\Delta\tilde{\mu}_{H^+}$ may be partly responsible for its increasing turnover rate but measurements indicate that any changes occurring are very small.

More importantly, a separate, novel, *control* system also exists. This involves a *regulator protein*, which binds to ATP synthase and inhibits it. At low respiration rates, most ATP synthase molecules are inhibited by this regulator, possibly preventing *hydrolysis* of cellular ATP at low $\Delta\tilde{\mu}_{H^+}$. As respiration speeds up, the regulatory protein is displaced and more synthase molecules are recruited for ATP synthesis. Factors which lead to ATP synthase activation through this system are $\Delta\tilde{\mu}_{H^+}$ and $[Ca^{2+}]$—thus again the enzyme is activated in times of energy demand and/or rapid respiration (Fig. 33.1c). The central role of Ca^{2+} ions in this regulation has been demonstrated by two approaches. First, treatment of the heart with *ruthenium red*, which blocks the Ca^{2+} uniport, will block the activation of these enzymes, leading to a fall in cellular [ATP]. Secondly, studies of cytoplasmic and intramitochondria free $[Ca^{2+}]$ using the fluorescent Ca^{2+} indicator *fura-2*, has demonstrated that Ca^{2+} ions do change in concentration as predicted by this model.

Another possible regulatory site is cytochrome aa_3. As noted above, this complex works slowly compared with the initial section of the respiratory chain, and must accelerate for faster respiration. This probably occurs simply by raising the concentration of its substrate, reduced cyt c (i.e. no separate regulation). However, ATP has been suggested as a possible (inhibitory) regulator (Fig. 33.1).

Control of mitochondrial respiration in liver

While the pathways of electron transfer and ATP synthesis are common to mitochondria of all tissues, control mechanisms will necessarily differ. The liver, for example, does not show wide variations in energy demand as does muscle, and its control will reflect this.

It appears that in liver:

1 The **substrate translocators** are important as potential control sites. This may reflect the ability of liver to switch from one fuel to another. However, apart from the fatty acyl transporter, and possibly the adenine nucleotide translocator, specific regulator molecules for the translocators are not known (Fig. 33.1).

2 The Ca^{2+}-linked system may be less important, and respiration modulated more simply by ADP levels (on the near-equilibrium model).

34 Uncoupling electron transfer from phosphorylation

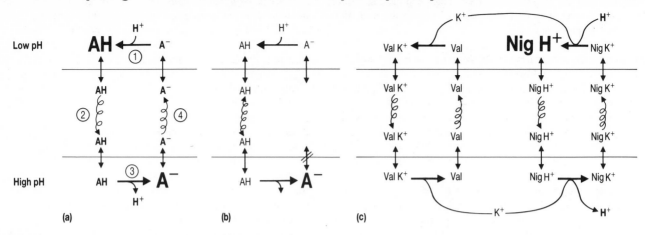

Fig. 34.1 Mechanism of uncoupling by lipophilic weak acids.
(a) Cyclic H$^+$ transfer by lipophilic weak acids (**uncoupling**).
(b) Accumulation of hydrophilic weak acid. Note that one molecule of uncoupler can move *many* H$^+$ ions, whereas normal weak acids can move only 1 mol H$^+$/mol acid.
(c) Cyclic H$^+$ transfer by combination of ionophores valinomycin (Val) and nigericin (Nig). Note that no net K$^+$ transfer occurs.

Inhibiting oxidative phosphorylation

Electron transfer is **coupled** to oxidative phosphorylation via a proton gradient; energy *released* from oxidations is *transferred*, via $\Delta\tilde{\mu}_{H^+}$, to the ATP synthase *trapping* system. The overall process is blocked by inhibitors like cyanide and antimycin which inhibit electron flow, DCCD and oligomycin which prevent the ATP synthase using $\Delta\tilde{\mu}_{H^+}$, and aurovertin and efrapeptin which inhibit the synthase itself (see Sections 12, 26). As the system is coupled together, a block at any point will inhibit the whole sequence—in particular, blocking the ATP synthase will slow electron flow (see Section 32).

A different class of compounds which block electron-transfer-linked phosphorylation are the **uncouplers**. These do not lead to the inhibition of all stages in the above sequence, but allow them to operate separately. In particular, they *allow electron transfer but dissipate the released energy as heat*. Uncouplers may attack the coupled system at *any* of the three stages (energy production, transfer or trapping), although a wide variety of chemical species affect the *energy transfer step*.

Ineffective gradient production—uncoupling the electron transfer complexes

Redox reactions may liberate free energy (ΔG negative). This energy would generally be dissipated as heat; however, biological systems have evolved to *trap* the energy from certain such reactions in a *proton gradient*. Whether energy is trapped or not depends on the *organization* of the electron carriers in the biological membrane, rather than on the reaction itself; the same reaction may produce proton transfer, or heat, depending on the organism involved. Examples already noted are the NADH/Q oxidoreductases of *E. coli* (heat producing; Section 14) or mammalian mitochondria (proton pumping; Section 11) and the two different terminal oxidases (one heat producing, one proton pumping) of plant mitochondria.

Thus it is, in principle, possible for a reagent to disrupt the organization of an electron transfer complex, preventing proton pumping while still allowing electrons to flow. Such a reagent would *uncouple* the electron transfer complex.

Not all respiratory complexes can be thus uncoupled, because in some cases H$^+$ transfer may be an integral part of electron transfer (QH$_2$ carries protons *and* electrons across the membrane, for

example). However, uncoupling can occur in complexes which employ redox *pumps* rather than proton translocating *loops* (Sections 24, 25). For example, mitochondrial cytochrome aa_3 is susceptible to the reagent DCCD which binds to its subunit III, allowing electron flow but decreasing proton pumping (Section 25). DCCD is thus an *uncoupler* of the cytochrome aa_3 complex.

Compounds with similar actions on the other respiratory enzymes have not been fully characterized. Such uncouplers of this type would be specific for one complex, requiring a specific interaction with it, unlike the protonophores, described below.

Dissipation of the proton gradient

Any chemicals, or manipulations, that allow protons to leak across membranes will uncouple, because they prevent the *transfer* of energy between electron flow and ATP synthesis. These include (i) compounds that disrupt membrane structure in general and (ii) compounds that specifically carry protons across the lipid bilayer (the **protonophores**). Surprisingly the mechanical disruption of membranes (e.g. by ultrasonic treatment) does not always lead to uncoupling. This is because the phospholipid assemblies produced tend to reseal into closed vesicles when the disruptive force is removed.

Agents which disrupt membranes

1 *Detergents* disrupt phospholipid/protein integration and are very powerful uncouplers.
2 *Phospholipases* create phospholipid fragments (lysophosphatides, fatty acids) which are powerful detergents.
3 *Pore forming proteins* e.g. mellitin, colicins, effectively punch holes in membranes.

Protonophores

These are agents that bind protons on the low pH side of the membrane, and carry them across. They are active in *catalytic amounts*—thus one uncoupler molecule deposits its proton on the high pH side and returns for another in a **continuous cycle** (Fig. 24.1a).

Four protonophoric uncouplers are shown in Fig. 34.2. They are not related structurally since they react with H$^+$ and the lipid bilayer, rather than with a specific protein. However, from Fig. 34.1a they all must have the following characteristics.

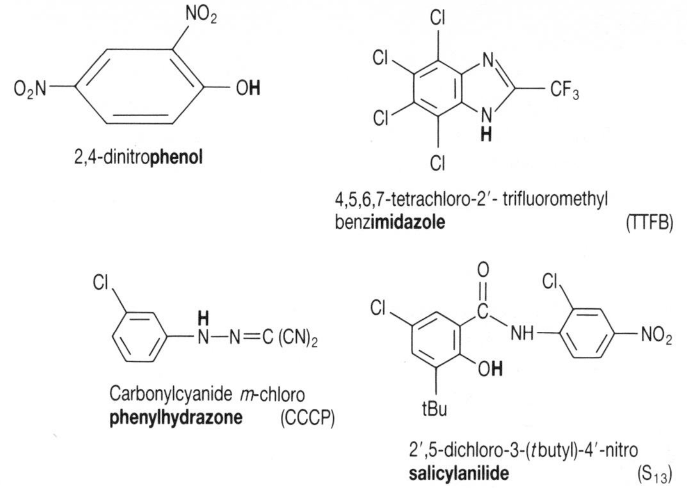

2,4-dinitrophenol

4,5,6,7-tetrachloro-2'- trifluoromethyl benzimidazole (TTFB)

Carbonylcyanide *m*-chloro phenylhydrazone (CCCP)

2',5-dichloro-3-(*t*butyl)-4'-nitro salicylanilide (S$_{13}$)

Fig. 34.2. Four uncouplers of oxidative phosphorylation. Classes of compound shown in **bold**. Other members of same class also uncouple. H = dissociating proton.

1 They are *acid-bases*, so they can release/pick up H^+ (steps 1,3);
2 They are *lipophilic*, so that both the *charged and uncharged* species can cross the membrane (steps 2,4).
3 They have a pK_a *between 5 and 9*, so that they ionize in the physiological range.

Note that weak acids such as acetic acid are not uncouplers. Acetate can bind H^+, and CH_3COOH can permeate bilayers—but the anion CH_3COO^- is too polar to cross the membrane and complete the cycle. Mitochondria will accumulate acetate, but are not uncoupled by it (Fig. 34.1b).

Ionophores

Compounds which carry ions other than H^+ should not uncouple electron transfer from phosphorylation. *Valinomycin*, for example, will carry K^+ across biological membranes. When applied to mitochondria, it will decrease $\Delta\psi$ by carrying positive charge inside; however, this simply results in ΔpH becoming larger (no charge difference opposing H^+ movement) and $\Delta\tilde{\mu}_{H^+}$ remains the same (Section 20).

Similarly, *nigericin*, which exchanges K^+ for H^+ across a membrane will not uncouple, it decreases ΔpH but increases $\Delta\psi$. However, a combination of these ionophores can produce a continuous cycle, carrying protons into mitochondria—and hence together they act as a protonophoric uncoupler (Fig. 34.1c).

H^+ channels

An *intrinsic transmembrane protein* which permits H^+ movement (H^+ uniport) (Section 31) will uncouple by dissipating $\Delta\tilde{\mu}_{H^+}$. Such proteins are operative in tissues important in heat production, such as brown adipose tissue in mammals (Section 37).

Uncoupling ATP synthesis from H^+ translocation through F_0

If F_1 structure is disrupted, protons may be able to cross F_0 but not interact productively with F_1. Agents which remove the smaller (δ, ε subunits) of F_1, abnormal (very low or very high) salt concentrations and some chemicals (e.g. orthophenylene dimaleimide) uncouple in this way.

The phosphate analogue, arsenate (AsO_4^{3-}, As_i), will substitute for phosphate in anhydride bond formation. Thus, energy from the proton gradient can be used to drive $ADP + AS_i \rightarrow ADP-AS_i$. This is a 'high energy compound' but, unlike ATP, it is *kinetically unstable* with a half life in water of less than 10 s. Thus, $ADP-As_i$ hydrolyses rapidly, with the result that $\Delta\tilde{\mu}_{H^+}$ is dissipated with no net ATP synthesis—uncoupling.

Like the uncouplers that act on the respiratory complexes, this action of As_i is *structure specific*. As_i binds to a specific P_i binding site on a protein. This is in contrast to uncouplers which carry protons (above). In addition, As_i (as a P_i analogue) also uncouples substrate level phosphorylation by an analogous mechanism (see Section 8).

Toxicity of uncouplers

Since uncouplers, in catalytic amounts, destroy the capacity for ATP synthesis, they are highly toxic compounds. The toxicity of arsenate is well known; the toxicity of the phenol uncouplers was discovered from their effects on workers in the explosives industry (manufacturing trinitrophenol, 'picric acid') early this century. Due to assimilation of dinitrophenol and similar compounds, these workers suffered severe weight loss and wasting of bodily mass— simply because they could oxidize organic material but the energy released was dissipated as heat.

Uncoupler resistant mutants

Uncouplers are also toxic to bacteria, destroying their essential transmembrane proton electrochemical gradient. At first sight, it might seem possible to select for uncoupler resistant mutants in an analogous way to selection for resistance to other antibiotics (e.g. streptomycin, tetracycline etc.) by growing bacteria on media containing protonophoric uncouplers (e.g. CCCP). In fact, only a very small number of resistant mutants have been isolated (from *Bacillus megaterium* and *E. coli*).

On further consideration, it is not surprising that so few protonophore resistant mutants are known, because these compounds react with H^+ and the phospholipid bilayer (above), not with any specific, mutable protein (as with antibiotics). Indeed, *it is surprising that any uncoupler-resistant mutants are known*; presumably any organism must retain a lipid bilayer membrane, which should allow protonophoric uncouplers to act. It might be that the uncouplers are destroyed or sequestered in the mutant strains, although as far as we can tell this is not always the case. The explanation for resistance to protonophoric uncouplers in mutant bacteria awaits further work; it is at present a mystery.

35 Human mitochondrial defects and disease

Symptoms of mitochondrial dysfunction

Since the major role of mitochondria is ATP synthesis, defects in mitochondria cause a deficiency of high energy phosphates within cells. This is particularly noticeable in tissues with a high requirement for energy. Thus skeletal muscle shows weakness (**myopathy**)—often noted in drooping eyelids (**ophthalmoplegia**)—as does heart muscle (**cardiomyopathy**), particularly under stress. The brain also malfunctions, leading to **encephalopathy** (linked to mental retardation) and/or **seizures**. In some cases all these systems are affected (*Kearns—Sayre* and related syndromes); however, symptoms may be restricted to one tissue such as brain (*Leigh's* syndrome) or skeletal muscle (*fatal infantile myopathy*). These symptoms are summarized in Fig. 35.1.

Defective mitochondrial ATP synthesis may arise from either a *block* in electron flow or ATP synthesis *or uncoupling* electron flow from ATP synthesis. In the former, oxidative metabolism is depressed, blood lactate levels are high (**lacticacidosis**) and fat droplets are deposited in muscle. In the latter, much rarer case (*Luft's* syndrome), hypermetabolism occurs; oxidation is rapid but inefficient in ATP production. In both cases, mitochondria (often deformed) will accumulate in affected tissues, in an apparent attempt to counterbalance the defect. This often gives a characteristic staining pattern of **ragged red fibres** in defective muscle, recognizable by light microscopy.

Measurement of ATP levels—in particular by the non-invasive technique of ^{31}P nmr—shows levels near normal in these cases. This may seem surprising, but it reflects the effectiveness of the normal controls on cellular [ATP]. However, these cases typically show *a markedly reduced creatine phosphate P_i ratio*—from $9:1$ in normals to as low as $1:1$ in some affected individuals. This is accompanied by a much *delayed recovery* in high energy phosphates after depletion by exercise.

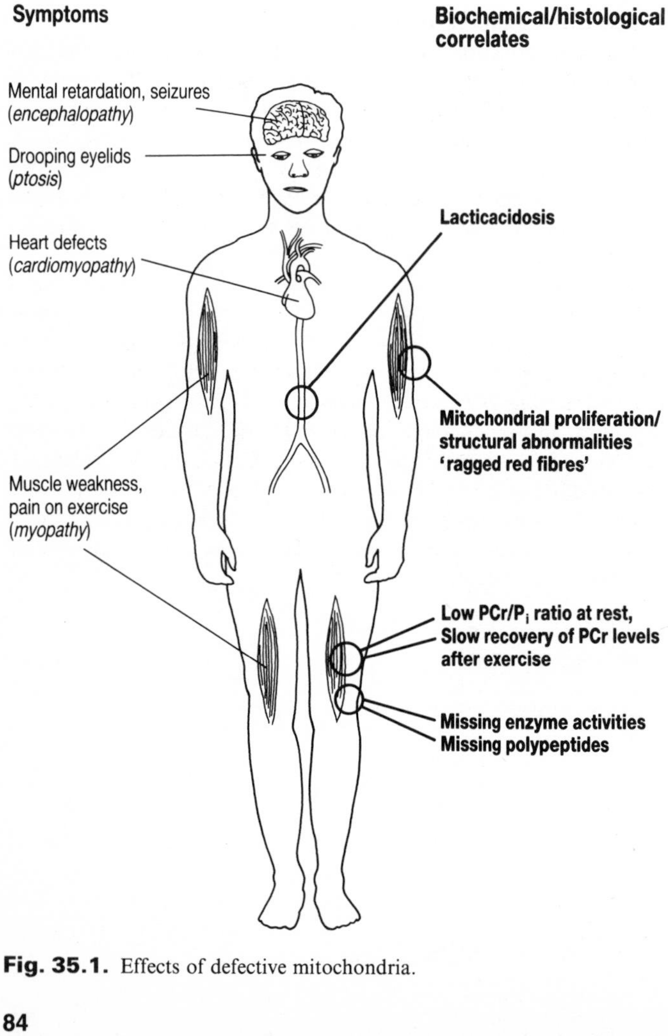

Causes of mitochondrial dysfunction

Mitochondrial assembly needs the cooperation of both the nuclear and the mitochondrial genome. Thus ATP synthesis may be blocked because (i) a mitochondrial gene is defective or missing, or (ii) a relevant nuclear gene is defective or missing. These are **primary defects** of the mitochondrion.

It is also possible for normal mitochondria to sustain damage *in vivo* and thus lose function—for example, in autoimmune disease (*primary biliary cirrhosis*), viral infection, poisoning (notably ethanol ingestion) or as a consequence of other malfunctions.* These are known as secondary mitochondrial defects, and will not be discussed here.

Defects in mitochondrial DNA

Human mitochondrial DNA (mtDNA) is just over 16 000 base pairs long, and codes 13 polypeptides in all, six of NADH dehydrogenase (ND1–6), one (cyt b) of the cytochrome bc_1 complex, three of cytochrome oxidase (CO1–3), and two of the ATP synthase (A6, A8). It also codes for those rRNAs and tRNAs required to translate these proteins (Fig. 35.2).

Some cases of mitochondrial dysfunction are caused by defects in mtDNA. A common class of defect involves a **deletion** of a considerable part (20–90 per cent) of the DNA, and thus total loss of several polypeptides. Interestingly, about half the observed cases have the *same* deletion, of about 5000 bases between bases 8470 and 13 460 (Fig. 35.2a). This deletion must have arisen independently several times, which indicates a deletion 'hotspot' in this region.

DNA sequencing shows the reason for this hotspot. The same 13 base sequence appears in mtDNA at bases 8470–8482 and 13 447–13 459; during replication these regions can, presumably, base pair and the DNA polymerase may slip from the earlier to the later region directly, omitting the intervening 5000 bases (Fig. 35.2b). Thus in mitochondria with this deletion, genes for two ATP synthase, one cytochrome oxidase and four NADH dehydrogenase subunits (and for several tRNAs) are absent, and the membrane is serious defective.

Other defects in mtDNA are known, where point mutations in the DNA occur. These may affect a single polypeptide, in which case their effect may be less severe than a large deletion. However, a number of point mutations in the tRNAs are know, where again a number of mitochondrial proteins are defective.

Deletion or mutation events occur very early during development, probably at oogenesis. Since they occur only rarely, only one or two mitochondria in the oocyte are likely to be affected and a cell is created with both normal and defective mitochondria. On division, the mitochondria segregate randomly to the daughter cells (**maternal or cytoplasmic inheritance**), and the resulting individual is a *mosaic* of cells with either normal or defective mitochondria (Fig. 35.2c).

The random nature of mitochondrial segregation may explain why patients with the same mtDNA deletion vary in the severity of their symptoms. It may also explain some variations in effects

Symptoms

- Mental retardation, seizures (*encephalopathy*)
- Drooping eyelids (*ptosis*)
- Heart defects (*cardiomyopathy*)
- Muscle weakness, pain on exercise (*myopathy*)

Biochemical/histological correlates

- Lacticacidosis
- Mitochondrial proliferation/structural abnormalities 'ragged red fibres'
- Low PCr/P$_i$ ratio at rest, Slow recovery of PCr levels after exercise
- Missing enzyme activities Missing polypeptides

Fig. 35.1. Effects of defective mitochondria.

* For example, in phenylketonuria, the abnormal metabolite phenylpyruvate can block pyruvate uptake, and thus oxidation, by mitochondria.

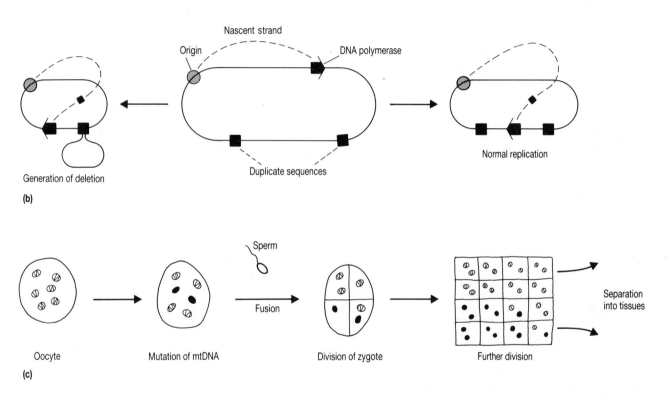

Fig. 35.2. (a) The human mitochondrial genome, showing the site of common deletion in mtDNA mutants. Note the genes for tRNA lying between the genes for polypeptides. (Numbers represent the number of base pairs from the replication origin.)

(b) Replication of mtDNA, generating a normal copy (right) or a deletion (left).

(c) Generation of a mosaic organism by random segregation of mtDNA defect.

between tissues. However, another factor here is rate of cell replication; muscle cells, which are laid down early in development, may retain the deletion while the population of leucocytes, which divide rapidly, might eliminate mutant mitochondria by somatic selection.

Defects in nuclear DNA

Nuclear DNA codes for the remaining polypeptides of the *respiratory chain*, *metabolite carriers* (for fatty acids, ADP etc.) and proteins of the *Krebs cycle*, together with other proteins of the mitochondrial matrix. It also codes for proteins needed for *import* of these proteins in mitochondria. Defects in proteins of all four classes are known.

Mutations in nuclear DNA (in contrast to those in mtDNA) normally lead to the loss of a single polypeptide. A number of respiratory chain defects are known; in one, for example, a defect in NADH dehydrogenase blocks electron flow (from flavin to quinone) (Section 12) and leads to fatal infantile myopathy (above). At the molecular level, a 75 kDa peptide, one of the iron–sulphur centres (as shown by epr spectroscopy), has been lost. Defects in several matrix enzymes too are known; in one case, a specific loss of lipoyl dehydrogenase inhibits both pyruvate and α-oxoglutarate dehydrogenases (see Section 33), and leads to cardiomyopathy and neural symptoms.

Tissue specific and developmental defects

Mutations in nuclear DNA (again unlike those in mtDNA) are inherited normally and will lead to cells with a uniform genotype. However, the *expressed defect* (phenotype) may be restricted to, say, skeletal muscle or brain (above). This is probably due to the existence of **tissue specific isoforms** of some respiratory chain polypeptides. For example, three of the smaller subunits of cytochrome oxidase, (VIa, VIIa and VIII) occur in at least two forms, the H (heart) and L (liver) forms (Fig. 35.3). Thus if only one of those isoforms is defective, the defect will be manifest only in that tissue.

A comparable situation may exist during development, in that fetal and adult isoforms of polypeptides may exist. In one case of *benign infantile myopathy*, cytochrome oxidase was deficient (6 per cent of normal) at birth but rose to normal levels, with disappearance of the symptoms, within 3 years. This is consistent with a defective fetal polypeptide being replaced by an adult form. It is

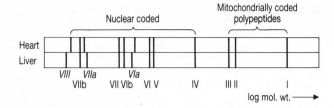

Fig. 35.3. Cytochrome oxidase polypeptides from liver and heart, shown as protein bands after electrophoresis in detergent. Isoforms produce bands of different mobilities in upper and lower traces (shown in *italic*). The (a), (b) designation is historical and does **not** relate to isoforms.

speculated that the reverse change, from normal fetal to defective adult isoform (e.g. of a respiratory chain complex) might be responsible for some cases of *sudden infant death syndrome* ('cot death').

Treatment of respiratory chain defects

Replacement of defective proteins, or even genes, is not yet possible in humans. Thus the treatment of respiratory chain defects has involved *bypassing the block with electron transfer mediators*. With a defect located in complex III, a *quinone* (menadione) plus *ascorbic acid* has proved effective in alleviating symptoms, presumably by transferring electrons from complex I directly to cytochrome *c* (Section 12). Similarly, *riboflavin* has been used to alleviate complex I deficiency.

36 Integration of chloroplast and cytoplasm

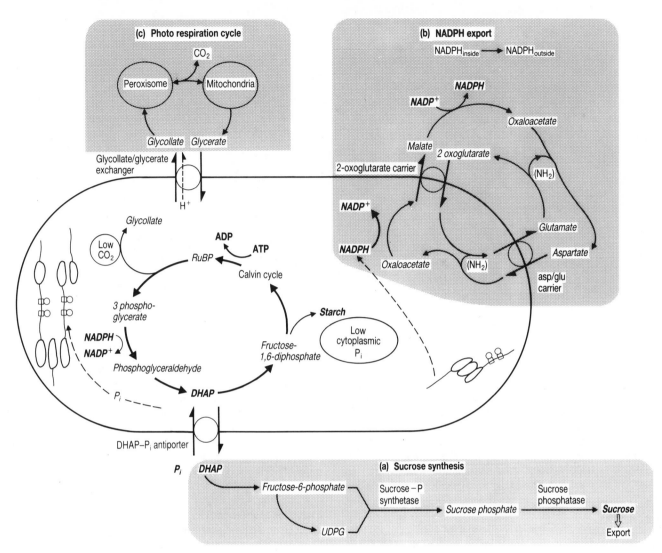

Fig. 36.1. Chloroplast anion carriers and interrelating systems.

Transporters of the chloroplast envelope

Like mitochondria, chloroplasts are bounded by an outer membrane, permeable to all small molecules, and an inner membrane impermeable to polar small molecules. Unlike mitochondria, however:

1 the energy transducing proton gradient is built up across a third internal membrane, the *thylakoid membrane* of the chloroplast (Section 17).

2 chloroplasts do *not* export ATP directly;

3 besides acting as a source of ATP, chloroplasts act as a source of fixed carbon and reducing power for biosyntheses.

As in mitochondria, the chloroplast inner membrane contains translocators (mainly antiporters) which allow communication between the stroma and cytoplasm (Fig. 36.1). The major carrier is an anion exchange protein which **exchanges cytoplasmic P_i for dihydroxyacetone phosphate (DHAP) formed in the stroma**. This triose phosphate can serve as a source of *carbon, reducing power and ATP* in the cytoplasm. Furthermore, it is raw material for the synthesis of sucrose, for export to other plant tissues (Fig. 36.1a).

The DHAP–P_i carrier polypeptide has a molecular mass of 29 kDa, with six (or seven) predicted transmembrane helices. (This arrangement is reminiscent of the mitochondrial anion carriers (see Section 31), but the two proteins do *not* appear to be homologous.) Its sixth transmembrane helix bears 2 intramembrane positive charges (lys 353, arg 354 in spinach), which may bind to the negative phosphate group during transport. This carrier is an **electroneutral** antiporter, and importantly, **conserves total stromal phosphate**; if free phosphate enters the stroma, phosphate is simultaneously lost as part of the triose phosphate.

Like mitochondria, chloroplasts also contain *dicarboxylate carriers*—in particular for malate/oxaloacetate and glutamate/aspartate, which take part in redox shuttles that export [H] from the chloroplasts (Fig. 36.1b) and, as in mitochondria, allow variations in redox state between the organelle and the cytoplasm. There are also carriers concerned specifically with carbon metabolism; chloroplasts from C_3 plants can export glycollate (made in photorespiration) for processing and recycling (Fig. 36.1c), and those from C_4 plants (e.g. tropical grasses) can exchange malate for pyruvate as part of the 'CO_2 pumping' mechanism of these plants.

Integration of carbon fixation and electron transfer

In the chloroplast, three processes must be linked together: *electron*

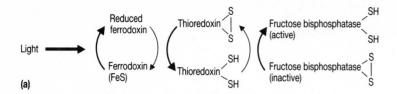

(a)

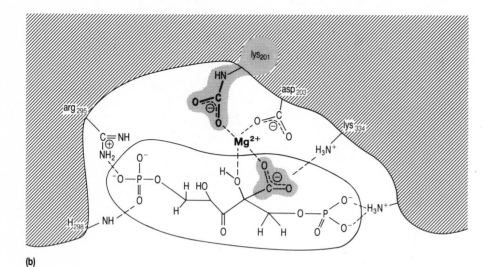

(b)

Fig. 36.2. (a) Activation of fructose biphosphatase by light.

(b) Active site of Rubisco, with bound carboxylated intermediate. Note the activating CO_2 bound to lys 201 (bold, shaded), the activating Mg^{2+} ion (bold) and the substrate CO_2 (shaded). Numbers refer to spinach sequence.

transfer (with concomitant ATP and NADPH production), *stromal carbon fixation* (in the Calvin cycle) and *export of triose phosphate* to the cytoplasm. In fact, cytoplasmic energy demand in plants is relatively constant, and the first two processes are not closely linked to the third. If the cytoplasmic $[P_i]$ rises (indicating a requirement for ATP, or production of sucrose (Fig. 36.1a) in the cytoplasm), the DHAP–P_i exchanger proceeds faster simply because P_i is a (non-saturating) substrate for the transporter. High levels of P_i also allosterically activate enzymes of the sucrose synthesis pathway. In contrast, if cytoplasmic P_i is low, CO_2 is still fixed—but the resulting carbon is *stored, temporarily, as starch within the stroma* and export is not required. Thus starch effectively acts as a carbon buffer, allowing the cytoplasm to receive a fairly constant flux of carbon, either from CO_2 fixation or from starch breakdown, in the light or in the dark.

'Substrate control' of photosynthesis within the chloroplast

Electron transfer and CO_2 reduction, on the other hand, are closely linked. Again, a simple model is one of 'substrate control'. At *low light intensities*, CO_2 fixation is limited by the availability of ATP and NADPH from electron transfer and just occurs as fast as it can. At *high light intensities*, ATP (and NADPH) build up because the supply of CO_2 becomes limiting to carbon fixation, and some degree of respiratory control operates—as ΔG_P rises, $\Delta \bar{\mu}_{H^+}$ rises and opposes the (proton pumping) electron flow (see Section 32). On this simple model, regulation of specific enzymes by regulator molecules (see Section 33) would not be required.

Although the plant can slow down electron flow (above) it cannot, of course, switch off incoming light energy if its intensity is too great. However, two mechanisms exist to limit absorbtion. First, if electron flow is blocked, energy is re-emitted as fluorescence (see Section 17). However, even the excitation of electrons by light is dangerous, promoting the formation of damaging free radicals. Thus, in addition, a mechanism of **photoinhibition**, which abolishes photochemistry altogether, is present (see below).

Over the long term, plants *adapt* to optimize the use of light and CO_2. Plants grown under low light conditions increase the amount of light harvesting chlorophyll per reaction centre to maximize the

chance of any photon producing electron flow (Section 16). The tropical grasses, habitually grown under high light conditions, have evolved the C_4 pathway to pump CO_2 into the CO_2 fixing chloroplasts (in 'bundle sheath' cells), and thus ensure that the maximum number of absorbed photons are useful. However, neither of these are acute regulatory mechanisms.

It is possible to consider photosynthesis as being substrate regulated in the short term, as indicated above. However, there are (as in mitochondria) individually controlled enzymes and thus true metabolic control mechanisms do operate.

Enzymes 'controlled' by light

Several 'non-equilibrium' enzymes of the Calvin cycle—*ribulose-5-phosphate kinase, fructose-1, 6-bisphosphatases, sedoheptulose bisphosphatase*—are 'switched on' only in the light. The same response is shown by the *ATP synthase*. If not switched off, these reactions would continue in the dark and the substrates of these enzymes would be severely depleted (ribulose-5-phosphate, fructose-1,6-bisphosphate, ATP etc.). When light returned, the Calvin cycle, lacking several intermediates, would proceed only very slowly. A graded control of these enzymes ensures that, at *low light intensities*, the rates of the enzymes of carbon fixation do not outstrip the supply of ATP and NADPH such that the cycle intermediates decline so far that fixation grinds to a halt.

Light affects enzymes indirectly in two ways. The first is via a reductase system; reducing equivalents generated by electron transfer are transferred from ferrodoxin to the protein, thioredoxin, which reduces S–S bridges on enzymes, converting them to the active dithiol $(-SH)_2$ forms (Fig. 36.2a). This unusual chemical modification is more suited to light regulated systems than the more familiar, kinase based systems. In the dark, reoxidation occurs (by O_2?) and the enzymes are inactivated.

A second mode of control occurs via ion movements. During photosynthetic electron transfer, H^+ ions are pumped *into* the thylakoid sacs. Thus (i) the pH of the stroma rises and (ii) the $[Mg^{2+}]$ content of the stroma rises, as Mg^{2+} ions are displaced from the thylakoids by H^+. A number of stromal enzymes (including Rubisco; see below) are *activated by Mg^{2+} ions and/or high pH*.

Problems in high light intensities

At high light intensities, in C_3 plants, the supply of reducing equivalents may exceed CO_2 supply. This may lead to (i) **photorespiration**, the 'fixation' of O_2 by **r**ibulose **bis**phosphate **c**arboxylase (**Rubisco**) in mistake for CO_2 (both compete for its active site), and (ii) **free radical damage** due to the reduction of O_2 (to superoxide) by the excess electrons (see Section 8). These problems are avoided by switching off both Rubisco and electron transfer at high light intensities.

Rubisco requires both Mg^{2+} and CO_2 for activity—both contribute to the structure of its active site (Fig. 36.2b). At zero electron transfer, Mg^{2+} ions are sequestered in the thylakoids, and Rubisco is inactive until they have been expelled into the stroma by the pH gradient (above). If sufficient CO_2 is present, this then *carbomoylates the active site lysine* (lys 201) and Rubisco is active.

At high light intensities, CO_2 fixation rates exceed CO_2 supply, CO_2 levels fall and the lysine is decarbamoylated; Rubisco is again inactive. This prevents O_2 occupying the active site of Rubisco, and hence inhibits photorespiration.

Electron transfer is also inhibited at high light intensities as $\Delta\tilde{\mu}_{H^+}$ builds up. As noted above, however, light absorbtion by chlorophyll still occurs, and this generates highly reducing electrons which are deleterious to the chloroplast (e.g. by reducing O_2 to superoxide). Chloroplasts counter this problem by **photoinhibition**, the light-dependent inactivation of photosystem II (PSII). Generation of $P680^+$, and QH· at too fast a rate for rereduction/reoxidation leads to oxidative damage (followed by *proteolysis) of the PSII polypeptide D1*, and thus a decrease in photochemistry to safe levels. This is a relatively expensive method of regulation, as a new PSII peptide must be synthesized each time—but it must be worthwhile within the overall economy of the plant.

37 Alternative uses of the proton gradient: 1 Heat generation

The importance of body temperature

Heat is a useful commodity in biological systems, because the rates of (bio)chemical reactions increase sharply with temperature. Multicellular organisms in temperate climate—even 'cold-blooded animals'—commonly function above ambient temperature so that their reactions can proceed at suitable rates.

Most of this heat comes simply from the operation of metabolism. Any metabolic process, even a biosynthesis, must be poised so that ΔG is negative overall, so as to maintain a flux (Section 4). Free energy released is dissipated as heat, and the organism's temperature is maintained. Additional heat may also be absorbed from the environment; basking snakes and lizards, for example, absorb heat directly from the sun's rays.

Uncoupling and heat production

If the above processes are insufficient to maintain temperature, an organism may oxidize substrates specifically to produce heat. Mitochondria perform these oxidations, but now the redox energy is not trapped as ATP—oxidation is uncoupled. Uncoupling can occur *in vivo* at any of the processes of *energy release, transfer or trapping* (Sections 19, 34).

1 *The redox chain may not pump H^+.* Plant mitochondria, in particular, possess non-pumping complexes (e.g. cyanide-insensitive Q oxidase) whose operation simply produces heat, as well as the customary energy conserving complexes. In some plants (e.g. *Arum* lily), this heat may be used to volatilize organic compounds to attract insects.

2 *The ATP produced may be rapidly and non-productively hydrolysed.* Actomyosin of mammalian skeletal muscle carries out this hydrolysis in shivering; simultaneous operation of phosphofructokinase and fructose-1, 6-bisphosphatase—a 'futile cycle'—does the same in insect flight muscle.

3 *Protons may be allowed to leak across the mitochondrial membrane.* This process occurs in brown adipose tissue, a tissue whose major role seems to be heat production in mammals. This mechanism is discussed further below.

Brown adipose tissue

Brown adipose tissue, as its name suggests, contains considerable amounts of stored triglyceride but, in contrast to white adipose tissue, it can be very active in oxidizing this fat. Thus

1 fat is stored in small droplets (rather than one large globule) to maximize surface area;

2 the tissue has a well developed blood supply;

3 the cells are rich in mitochondria, with well developed cristae. (The cytochromes etc. are responsible for the 'brown' colour.)

In hibernating animals such as the ground squirrel, the body temperature drops to 10°C on hibernation.* On arousal, this rises rapidly (within 5 min) to 35°C due to fat oxidation in brown adipose tissue. Over 60 per cent of bodily oxygen consumption can ·be switched to this tissue by noradrenergic nervous stimulation and used for heat production.

In non-hibernating animals, the amount of brown adipose tissue is generally small. However, it is higher in *newly born* or *cold adapted* animals, again suggesting a role for this tissue in heat generation. Furthermore, an inherited defect in brown adipose tissue in mice (specifically in the uncoupling protein—see below) leads to cold intolerance, but also to *obesity*. By extrapolation, it has been sug-

gested that brown adipose tissue may be able to 'burn off' excess food in humans, and hence regulate body weight. While still speculative, this idea has attracted much interest in the field of human obesity and diet.

The uncoupling protein

Mammalian mitochondria from tissues other than brown adipose tissue are relatively impermeable to protons at normal levels of $\Delta\tilde{\mu}_{H^+}$ (Section 20). At very high levels ($\Delta\psi > 220\,mV$), leakage becomes significant ('dielectric breakdown'), possibly to protect membranes against adverse effects of high voltage gradients (Fig. 37.1, right hand curve).

Brown adipose tissue mitochondria contains the H^+-pumping respiratory chain typical of mammalian mitochondria, and a normal H^+-driven ATP synthase. However, in addition, protons can cross the membrane through a tissue-specific channel, the **uncoupling protein**. This channel can be controlled; when it is closed, H^+ permeability of the membranes is small (below $\Delta\tilde{\mu}_{H^+} = 220\,mV$, as above). However, when heat is to be produced, the channel is opened and H^+ leakage becomes significant in the physiological range of $\Delta\psi = 160$–$220\,mV$ (Fig. 37.1, left hand curve), and uncoupling occurs.

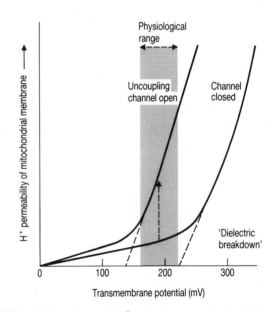

Fig. 37.1. $\Delta\psi$ dependence of H^+ leakage across mitochondrial membranes. The dotted arrow indicates the effect of opening H^+ channels on proton permeability at $\Delta\psi = 200\,mV$.

The uncoupling protein is, surprisingly, homologous to the adenine nucleotide carrier (and other anion transporters) of mitochondria. Thus it consists of three regions of about 100 amino acids, each homologous with the other two and comprising 2 transmembrane helices and an intervening loop. The structure of such proteins is described in detail in Section 31. Here, we should note that H^+ is very simple in structure and positively charged; it is thus surprising to find it carried on *transporters* for complicated molecules like ATP, which are negatively charged, and not on *channels* like the DCCD-sensitive H^+ channel of the ATP synthase (Section 30). One possibility is that the uncoupling protein may carry protons as H_3O^+, or OH^- in the opposite direction. In this case it will require chelating groups like –C=O and –NH, rather than specific acid–base groups (like COOH), in the membrane phase.

---

* Not all hibernating animals show this temperature drop. Bears have a body temperature above 30°C even when hibernating.

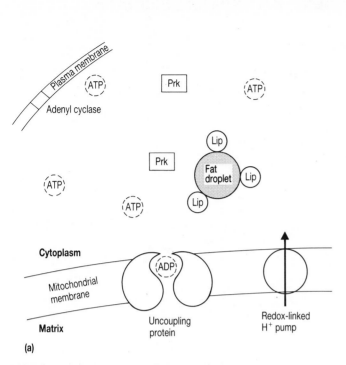

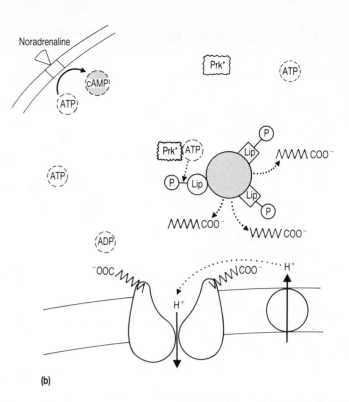

Fig. 37.2. Activation of H$^+$ channel in brown adipose tissue mitochondria. Lip = lipase, /\/\/\/\/\ COO$^-$ = fatty acid.
(a) Before activation.
(b) After activation.

Control of proton permeability

The activity of the uncoupling protein is regulated by **purine nucleotides** (GDP, ADP) and **fatty acids**. In the absence of GDP or ADP, the channel is open, and proton permeability high. The diphosphates (probably ADP *in vivo*) bind to the protein, and close the channel. Fatty acids, on the other hand, increase H$^+$ permeability—again by binding to the uncoupling protein.

Levels of nucleotides *in vivo* are such that the nucleotide binding site is occupied, and proton permeability low. (It is unlikely that nucleotide levels ever change sufficiently to regulate channel activity *in vivo*.) In the body, the channel opens in response to *neural stimulation* of brown adipose tissue, releasing noradrenaline from the sympathetic nerve terminals. This hormone activates a lipase (via the cAMP cascade), levels of fatty acids rise, and *the proton leak is activated* (Fig. 37.2). At the same time, *the fatty acids provide a supply of an oxidizable substrate* to fuel heat production.

The above describes acute regulation of heat production by brown adipose mitochondria (e.g. on arousal from hibernation). In cold adaptation of rats, for example, chronic regulation is also seen. In particular, protein synthesis leads to the incorporation of uncoupling protein into existing mitochondrial membranes.

Measurements of H$^+$ permeability: mitochondrial swelling

H$^+$ permeability of mitochondria can be followed in O$_2$ pulse experiments (Section 20). Brown adiposes tissue mitochondria show normal H$^+$ extrusion (via their redox-linked pumps) but, when the H$^+$ channel is open, the protons rapidly return across the membrane.

A more convenient, if less direct, indication of H$^+$ permeability is *(passive) mitochondrial swelling*. This can be measured as changes in light scattering in a mitochondrial suspension—as mitochondria swell, light scattering decreases.

Mitochondria swell (due to osmotic water intake) if they accumulate salts. If mitochondria are in osmotic balance with their surroundings, swelling can be induced by adding *both a permeant cation and anion*—permeation of only one would cause a charge imbalance, preventing further uptake. For example, mitochondria suspended in potassium thiocyanate (KCNS) will not swell, because CNS$^-$ is permeant but K$^+$ is not; they will swell if valinomycin is present, for then K$^+$ becomes permeant too (Section 34).

To investigate the proton channel, mitochondria are suspended in potassium acetate + valinomycin; they will not swell because the anion CH$_3$COO$^-$ (Ac$^-$) is impermeant. However, Ac$^-$ is a weak base—it can pick up a proton and cross the membrane as HAc (Section 34). Still, the charge imbalance persists and no swelling occurs. If, now, the membrane is made permeable to protons e.g. by activating the uncoupling protein, H$^+$ can move out, counteracting the charge imbalance. K$^+$Ac$^-$ can now be accumulated, *and the mitochondria swell.*

38 Alternative uses of the proton gradient: 2 Transport systems

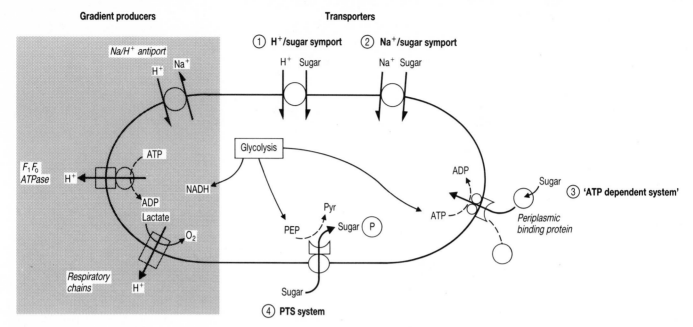

Fig. 38.1. Four classes of sugar transport system in *E. coli*. Examples are Type 1 H^+/lactose, arabinose, xylose, rhamnose. Type 2 Na^+/mellibiose.

Type 3 ATP/maltose, ribose, arabinose (plus amino acids his, gly, arg). Type 4 PEP/fructose, glucose, sucrose, mannose.

Modes of transmembrane transport

To accumulate hydrophilic compounds or ions, biological systems require (i) a *transmembrane protein*, to ferry the molecule through the lipid phase, and (ii) an *energy source*—either ATP or an ion gradient.

In **eukaryotic cells**, Na^+ is pumped out of the cell using ATP (Section 42), while amino acids, sugars etc. are accumulated across the plasma membrane using energy from this Na^+ gradient. Mitochondria, which pump out H^+ via electron transfer, use their H^+ gradient to make ATP and to accumulate P_i. Both their H^+ and P_i gradients are used to import other anions (Section 31). These relationships emphasize *the interchangeability of energy between ATP and ion gradients*, and *the ability of any ion gradient to act as an energy source.*

In **bacterial cells**, this interchangeability is even more striking. For sugar import alone, *E. coli* has available four classes of transport system (Fig. 38.1).

1 H^+-dependent systems, driven by the transmembrane $\Delta\tilde{\mu}_{H^+}$.

2 Na^+-dependent systems, using $\Delta\tilde{\mu}_{Na^+}$ (which in turn is built up using $\Delta\tilde{\mu}_{H^+}$).

3 ATP-dependent systems, which employ a soluble, *periplasmic binding protein* to bind the sugar in addition to the transmembrane carrier protein.

4 Phosphoenol pyruvate (PEP) utilizing systems, which use PEP both as a source of energy for transport *and* to phosphorylate the sugar. These systems are known as the *phosphotransferase* (PTS) systems.

All four systems are depicted in Fig. 38.1, together with their sources of energy. Within each class, there are a number of individual transporters: ATP driven systems drive maltose, arabinose, histidine etc. into *E. coli*; PTS systems drive glucose, fructose, sucrose etc. Clearly, these are not always expressed in the bacterium, but are *inducible* if the relevant compound is present.

Another factor affecting the use of a particular transport system by a bacterium is the external *concentration* of the compound to be taken up. For example, there exist both H^+-dependent and ATP-dependent transporters for arabinose in *E. coli*. Movement of 1 H^+ down its electrochemical gradient typically yields 15–20 kJ; hydrolysis of ATP 55–60 kJ. This means that, energetically, the *accumulation ratio* (concentration inside/concentration outside) using ATP can be much larger (10^7–10^9) than using H^+ (10^3–10^4). Thus ATP driven systems tend to appear when the external concentration of the compound required drops below about 10^{-6} M (1 μM).

ATP driven transport systems are dealt with in Sections 42–44; this section concentrates on the H^+-dependent transport systems of bacteria, and their use of the transmembrane H^+ gradient to drive transport.

H^+-dependent systems

The H^+-dependent transporters in bacteria exhibit a common structure. This comprises a single polypeptide with 12 transmembrane helices (Fig. 38.2).

Sugar transport derives its energy from $\Delta\tilde{\mu}_{H^+}$ via a **symport system**. *The sugar*, moving up its potential gradient, *moves into the cell together with H^+* (down its potential gradient) so that the overall process has a net negative ΔG.* For tight coupling (i.e. to prevent leakage), movement of either component individually is prevented.

Such a system needs

1 a sugar binding site

2 a reorientating pore to allow the sugar across the membrane (see Section 44)

3 a H^+ binding site.

---

*This can be contrasted with the *Na^+/H^+ antiport system* shown in Fig. 38.1. Here Na^+ moves uphill and H^+ moves downhill, but now movement is in *opposite directions*. The energetics is unchanged—but the mechanism is adapted for a different purpose.

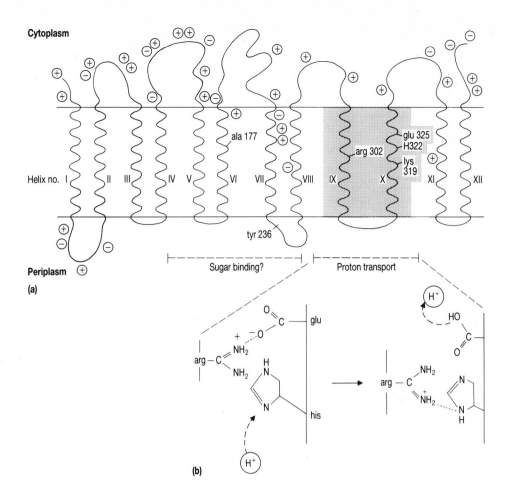

Fig. 38.2. (a) Two-dimensional representation of lactose symporter, showing putative functional amino acids.
(b) 'Relay' system for H$^+$ transfer.

Our two-dimensional models (Fig. 38.2) do not allow us to identify the pore. It may be that the hydrophilic sides of some transmembrane helices line a pore, and that the helices tilt relative to each other when reorientating, but this is a speculative model.

The lactose carrier

Lactose enters *E. coli* cells on the **H$^+$/lactose symporter** (also known as *lactose permease*). A two-dimensional model for this protein is given in Fig. 38.2; like all carriers, we do not know the three-dimensional structure as yet. This model has been used as a basis for studies on the roles of individual amino acids in the function of this protein.

One aim is to identify the H$^+$-*binding site* of the symport. This involves (i) inspection of the sequence to identify acid–base groups within the membrane phase and (ii) modifying such side chains by site directed mutagenesis, followed by activity measurements on the modified protein. These methods indicate that his 322 and glu 325 on helix X and arg 302 on helix IX may participate in H$^+$ movement, possibly in a 'proton relay' system (Fig. 38.2b; cf. Fig. 40.2). Conversion of glu 325, for example, leads to a protein capable of transporting lactose (downhill) but not of H$^+$ driven transport, while modification of other histidine residues or the nearby ser 300 or lys 319, has no effect.

A similar approach has implicated ala 177 (helix VI) and tyr 236 (membrane surface) in sugar binding; changes of ala 177 to val, for example, lead to an increase in ability of the carrier to transport maltose. However, our picture of this site is far from complete.

Other H$^+$/sugar symporters

Amino acid sequences are known for a number of other bacterial H$^+$/sugar transporters. All comprise 12 (putative) transmembrane helices, like the H$^+$/lactose symporter. Among these symporters can be recognized several families, which are clearly related to each other by sequence homology (typically > 33% identity between amino acid sequences). Thus the *H$^+$/lactose symporter* is related to the *H$^+$/sucrose symporter* of *E. coli*. Another family contains carriers for arabinose, xylose and galactose.

Although less functional information is available for the arabinose family of carriers, they cannot employ a similar mechanism to the lactose carrier. In particular, there are no conserved acid–base groups in helices IX and X; the only such conserved intramembrane group is asp 39. Different families of carriers may thus use different detailed mechanisms to accomplish the same symport function.

Homologies between bacterial and mammalian transport systems

Homologies between a subset of sugar/H$^+$ symports in *E. coli* may not be surprising. However, protein databases reveal homologies between the arabinose family of bacterial transporters and the *glucose transporter of mammalian plasma membranes* including the insulin-dependent transporter of muscle. There are also functional similarities—all are *inhibited by cytochalasin B*.

These similarities are remarkable not only because of the evolutionary distance between *E. coli* and mammals, but also because of a *fundamental difference in energetics*. The mammalian glucose transporter mediates **facilitated diffusion** of glucose down a concentration gradient without any accompanying ion—it is a *uniport*. It thus appears that modification of a carrier from a downhill to an energy driven symport may require a relatively minor change. (Support for this idea in principle comes from mutagenesis of the H$^+$/lactose symport to a downhill lactose carrier, mentioned above.) Interestingly, the glucose carrier does not bear asp 39 in helix I, which was suggested as a site of H^+ binding in this family of transporters.

A 'superfamily' of 12 helix transporters?

The lactose and sucrose symporters of *E. coli* are related by sequence, as are the arabinose, xylase and galactose symporters and the mammalian glucose transporters. More detailed comparisons between these sequences suggest that these two families may in fact have diverged from a single common ancestor; the two families are, in turn, members of a **superfamily**. This superfamily contains three other families of 12 helix transporters, those containing: (i) the citrate/H^+ symporter; (ii) the glycerol-3-phosphate/P_i antiporter; and (iii) the H^+/tetracyline antiporter, all from *E. coli* and all, presumably, derived from a common ancestor. Thus, variations on a common theme seem to allow for a surprisingly wide variety of transport mechanisms and specificities, making it all the more tantalizing that we do not know the mechanism or structure of *any* in detail.

The general theme of 12 transmembrane helices among transport proteins is also seen in apparently unrelated proteins—the mitochondrial transporters (Section 31), the PTS transporters and the ATP driven bacterial transporters (above). The last mentioned category also have mammalian relatives. The significance of this is not known—presumably 12 transmembrane helices are, for some reason, a good framework on which to build a transporter.

39 Alternative uses of the proton gradient: 3 Bacterial motion

Directed motion in bacteria

Bacteria 'swim' through liquid media either to approach food sources (positive chemotaxis), or to avoid unfavourable environments (negative chemotaxis). *E. coli*, for example, can move at about 50 µm (about 20 times its body length) per second. To do this, they employ a *motor* to provide forward thrust—the **flagellum**—powered by an energy source. $\Delta\tilde{\mu}_{H^+}$. Some bacteria like *E. coli*, possess several (6–8) flagella per cell body; others, like *R. rubrum*, have only one.

Flagella can drive movement only along a line. Directionality in bacterial swimming is achieved in a series of *punctuated runs*. The bacterium swims in one direction for some time and then stops; Brownian motion will then *randomly reorientate the organism* (**tumbling**), *which will now set off in a different direction. The bacterium has no control over the reorientation.*

Chemotaxis biases direction by increasing the length of run between tumbles (if movement is in a favourable direction) or decreasing it (if unfavourable) (Fig. 39.1). This, of course, requires the bacteria to have systems for *sensing* attractants etc. which interact with the switch mechanism (below). These systems are not discussed here.

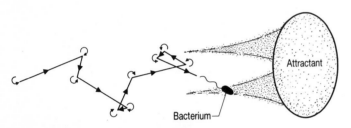

Fig. 39.1. Movement of bacterium towards an attractant. When the bacterium is swimming towards the attractant, it tumbles infrequently; when it is swimming away, it tumbles more frequently. This biases motion towards the attractant.

Structure of the flagellum

The structure of the flagellum from *E. coli* is shown in Fig. 39.2. It is made up of the **basal region** (which lies within the membrane and cell wall), the **hook** and the **filament** (which protrudes into the medium up to about 10 µm). The whole structure contains the products of up to 35 genes (designated *fli* or *mot*). Most of the polypeptides lie in the basal region, which appears as a complex 'layered' structure in the

electron microscope (Fig. 39.2). The filament contains only a single protein (**flagellin**) and the hook another.

The arrangement of these proteins is not known in detail. However, since the flagellum *rotates anticlockwise*, and provides *forward thrust*, mechanical considerations require:

1 A static anchor (**stator**) within the membrane (possibly anchored to the cell wall) at which torque can be generated. This is the product of (at least) genes *mot A* and *mot B* which form a circle of 8–12 subunits around the rotor.

2 A **rotor**, which pushes against this anchor to cause rotation. This must also contain the on/off (or clockwise/anticlockwise in *E. coli*) *switch* which causes tumbling. This is the product of genes *Fli F* (the major structural component) and *Fli G, M* and *N*, which make up the switch mechanism.

Together, **1** and **2** comprise the *S and M rings* as seen by electron microscopy of flagella isolated by osmotic shock.

3 A **bush** to provide frictionless passage of the flagellum through the cell wall and (in the case of gram negative bacteria) the outer membrane (*L and P rings*).

4 A *left-handed helical array of flagellin subunits* in the **filament**. Left-handed geometry is important, in order that anticlockwise *rotation* provides forward *thrust*, like a rotating screw. Typically there are some 20 000 subunits in one filament and these are organized into eleven protofilaments arranged in a cylinder with a 6 nm central hole. Two of the protofilaments contain flagellin in an unusual conformation; these are slightly shorter than the other nine, and contribute to the screw twist of the flagellum.

Other components of the system are the **hook** (whose specific function is unknown), and various proteins that operate the switch on the rotor in response to external stimuli. These latter are products of the *che* (chemotaxis) genes.

Energy transduction in the flagellum

The energy to drive flagellar rotation comes directly from the transmembrane proton gradient ($\Delta\tilde{\mu}_{H^+}$ ATP is not needed to power rotation. Thus, in F_1, F_o deficient mutants, electron transfer can drive rotation. The force produced is proportional to $\Delta\tilde{\mu}_{H^+}$ (experimentally, the rate of movement is proportional to $\Delta\tilde{\mu}_{H^+}$) and independent of the resistance applied (experimentally, the rate of movement is inversely proportional to viscosity). This suggests that there is a *fixed stoichiometry of H^+ moved per rotation*. This is reminiscent of the ATP synthase, where H^+/P, the number of protons required to make 1 ATP, is a fixed value of 3, irrespective of the thermodynamically most efficient value (Section 21). However,

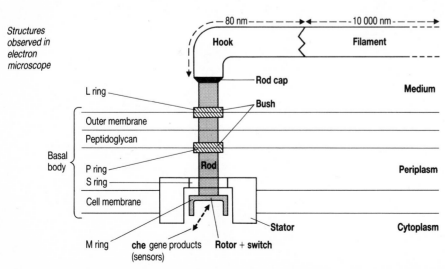

Fig. 39.2. Structure of the *E. coli* flagellum, showing observable structures.

in the case of the flagellum, *several hundred (100 < n < 1000) protons are required for a single rotation.*

A model for flagellar rotation

A tentative model for energy coupling in the flagellum is shown in Fig. 39.3.

1 H^+ binds to a basic group on the outside of a stator protein (high $\Delta\tilde{\mu}_{H^+}$ outside).

2 This protein changes in conformation, and dissociates from one subunit of the rotor.

3 The conformation change allows the proton access to the (low potential) inside of the cell, and H^+ dissociates inside. The *mot A* protein provides the H^+ channel through the membrane.

4 The stator protein associates with the next rotor subunit.

5 The stator protein relaxes to its original conformation, pulling the rotor around. The cycle can then repeat.

This scheme incorporates the essentials of energy transduction—movement of H^+ (or possibly H_3O^+) is *gated* and *proton binding, from a high potential, is used to weaken binding between two favourably associated proteins*—a conformational coupling mechanism. (It is, indeed, a combination between models for the ATP synthase (Section 28) and actomyosin in muscle.) This mechanism can also incorporate Na^+ substitution for H^+, as occurs in some marine bacteria (e.g. *Vibrio alginolyticus*), by altering the nature of the ion binding site in the first step.

On this model, the number of subunits in the rotor would be n', the number of protons required per revolution or, if say 3 protons were gated per conformational change, the number in the rotor, would be $n'/3$. Unfortunately, our structural knowledge of this system is so small that even the existence of a multisubunit rotor is unproven.

This model is compatible with the kinetic properties of the system—replacing H_2O by D_2O does not influence the rate of motion, suggesting that H^+ dissociation rate is fast relative to other steps. However, since this model depends on protein conformational changes, we might expect rotation rate to be temperature dependent—but it is not. The reason for this is unclear.

Evidence for rotation

It was stated above that the bacterial flagellum rotates. This is very surprising—eukaryotic flagella do not rotate, but provide thrust by bending ('whiplash'). Indeed, no other biomolecular system is known to rotate, making the bacterial flagellum unique in evolution.

We cannot observe the rotation of flagella directly. Experiments which prove that rotation does indeed occur are as follows:

1 Flagellated bacteria stick to microscope slides coated with antiflagellin antibody. In this case, the flagellum is *tethered*, and cannot move. In this case the large cell body can clearly be seen to rotate.

2 Small latex beads can be attached to the flagellum, and rotation can be observed microscopically.

These techniques can also be used to show that a helical flagellum is required to develop thrust (above). Some non-motile mutants—'straight' mutants—do not show a helical arrangement. The above techniques show that their flagella do still rotate; they are non-motile because no thrust develops in the absence of a helical organization.

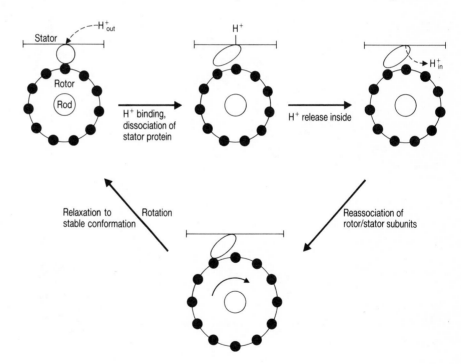

Fig. 39.3. A possible model for the conversion of $\Delta\tilde{\mu}_{H^+}$ into flagellar rotation.

Direct transduction of light to gradient energy

In biology, proton pumping is typically powered by ATP hydrolysis or electron transfer. Even in chloroplasts, protons are pumped by redox reactions in an electron transfer chain; light energy is used merely to feed electrons into this chain (Section 16).

Certain organisms, notably the archaebacterium *Halobacterium halobium*, can, in contrast, *transduce light energy directly into a proton gradient* without intervening redox steps. The transducer is the purple transmembrane protein **bacteriorhodopsin**. When *H. halobium* lacks oxygen and/or organic substrate, this protein is found in high concentration (effectively a two-dimensional crystal) in specialized regions of its cell membrane, the *purple patches.**

This perhaps rather obscure protein has been extensively studied for several reasons:

1 Its alignment in a *crystalline array*, in the native membrane, without other proteins present, makes it favourable for structural studies by electron diffraction.

2 As a *single polypeptide* capable of transducing light energy into a proton gradient, it is a relatively simple system (a 'molecular machine') on which to study energy transduction.

3 It is structurally analogous to the *visual rhodopsins* and its study has shed light on the structure of, and signal transduction by, these pigments.

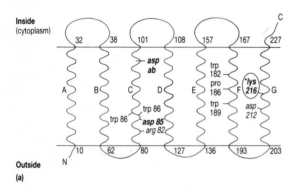

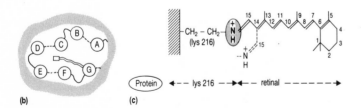

Fig. 40.1. **(a)** Two-dimensional map of bacteriorhodopsin. Residues in H$^+$ channel shown in *italic*, with key acid–base groups in **bold italics**. Other residues shown lie in retinal binding pocket. Numbers indicate amino acids at ends of transmembrane helices.

(b) View of bacteriorhodopsin from outside membrane surface, showing helix connectivity (dotted line = loop on cytoplasmic surface) and orientation of retinal (shaded region = surrounding lipid).

(c) Retinal binding to lys 216, with the protonated Schiff's base shown in bold. The all *trans* form is shown in solid lines, the 13-cis form is dotted.

*At high oxygen tension, the organism possesses a typical respiratory chain and survives by oxidizing organic material, like other bacteria.

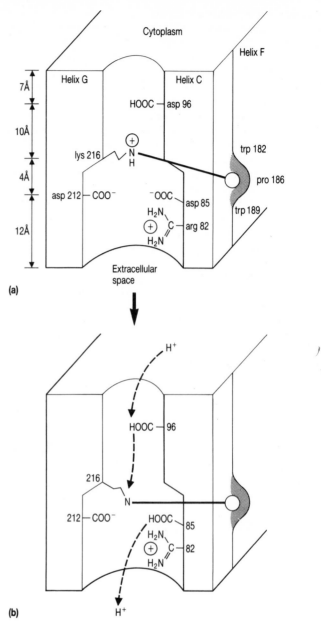

Fig. 40.2. Schematic view of bacteriorhodopsin active site, showing wide (hydrophilic) channel on outside, and narrow (hydrophobic) channel on the cytoplasmic side.

(a) Resting state.

(b) Excited state, showing deprotonated Schiff's base and reorientation of the retinal.

Structure of bacteriorhodopsin

Bacteriorhodopsin comprises *seven transmembrane helices*, A–G, oriented *nearly perpendicular* to the membrane surface (Fig. 40.1a). Its N-terminus lies on the extracellular surface of the cell. These helices associate in a horseshoe-shaped tertiary array, connected sequentially around the horseshoe (Fig. 40.1b). Association between helices relies on salt bridges between their occasional charged side chains, and association between protein and surrounding lipid involves the hydrophobic side chains.

The chromophoric group, **retinal** (which is responsible for the absorption of light), is attached to the protein covalently via a *Schiff's base* (Fig. 40.1c) between its aldehyde group and **lys 216** (helix G). It is situated roughly in the centre of the membrane phase, nearly, but not quite, parallel to the membrane surface. The Schiff's base is close to two negatively charged aspartate groups (asp 85, asp 212) and is normally protonated. The remainder of the retinal molecule fits into a hydrophobic 'pocket' involving residues from various helices—notably tryptophan and tyrosine residues from helices F and C.

On the extracellular side ('below' in Fig. 40.1a), the Schiff's base communicates with the medium via a *water-filled cavity* containing not only asp 85 and 212 but also various hydrophilic (tyr, arg, glu) residues—notably arg 82 which interacts with asp 85. On the cytoplasmic side ('above' in Fig. 40.2), there is a single hydrophilic residue, asp 96, and a number of hydrophobic residues (Fig. 40.2); access of water to asp 96 is probably limited.

Mechanism of H⁺ pumping

On the basis of this structure (derived largely from electron diffraction of frozen specimens) and by correlation with functional studies (where important residues are modified by site-directed mutagenesis), the following model is proposed to describe proton pumping by bacteriorhodopsin. Essentially, the proton is carried by the Schiff's base $-\overset{+}{N}H-$; light causes this to release H^+ outside the cell (via asp 85) and, after a reorientation of the base, to take up H^+ from inside (via asp 96).

1 In the non-excited state, retinal is in its all *trans* form, and is located centrally in the membrane. The Schiff's base is attached at lys 216 (helix G); the remainder of the chromophore spans the horseshoe with its distal hydrocarbon ring in the hydrophobic pocket between helices C and F. *The Schiff's base is protonated* ($pK_a \approx 10$), as is asp 96; asp 85 and asp 212 are ionized, close to the positively charged arg 82 (Fig. 40.2a).

2 On absorbtion of a photon (≈ 200 kJ/mol) *the retinal isomerizes to the 13-cis form*, with an abrupt change in geometry (Fig. 40.1c). Its distal, hydrophobic end is rigidly held in its pocket, so the *-NH- group moves* towards the outside of the cell (downwards, in Fig. 40.2). This lowers its pK_a and the *proton is transferred to asp 85*.

3 A *conformational change* occurs, involving a bending of helix F at pro 186. This allows *asp 85 to release its proton* to the outside medium (preventing reverse transfer back to the Schiff's base). It also allows *asp 96 to transfer its proton* onto the base, restoring it to the $-\overset{+}{N}H-$ state (Fig. 40.2b).

4 Asp 96 binds H^+ *from the cytoplasmic aqueous phase*, and returns to the uncharged state. Both the protein and the retinal then *relax* to their original conformational state, ready for the cycle to begin again.

Note that this mechanism requires only *very small changes in the protein conformation* during proton pumping; most of the process is mediated by conformational changes (*trans* to *cis* isomerization) around one bond to the retinal moiety.

The conformational changes, and protonation changes, in this cycle can be followed spectroscopically (Fig. 40.3), the varying conformational states of retinal having absorption in the region 550–600 nm when protonated, and around 400 nm when deprotonated. This allows us to follow the various stages of the process in time. Isomerization of retinal occurs within 10 ps, as one would expect for a photochemical process. Subsequent deprotonation, through the wide, water-filled channel, takes about 50 μs. Reprotonation involves proton entry into the hydrophobic end of the pore, is slower (10 ms), and limits the reverse isomerization of the retinal and thus the turnover of the system.

Homologous systems

Halobacterium contains two further pigments which are homologous (about 35 per cent sequence homology) to bacteriorhodopsin—a chloride pump, **halorhodopsin**, and a sensory pigment, **sensory rhodopsin**. Both contain retinal, and have the same basic structure as above. Their specificities may be governed, at least partially, by residues in the upper (asp 96) channel of the molecule. In the case of halorhodopsin, this residue is absent and the whole channel is lined with smaller residues, making a larger, less hydrophobic hole (or Cl⁻ passage). Sensory rhodopsin again lacks the equivalent to asp 96, but has this channel blocked by bulky residues. In signalling, the conformational change induced by light is presumably detected directly by a transducer protein.

Structure of bacteriorhodopsin—a paradigm for membrane protein structure

Bacteriorhodopsin was the first membrane protein whose structure was defined. From it, we learnt that:

1 a stretch of some 20 hydrophobic amino acids is required to traverse the lipid bilayer of a membrane,

2 these amino acids form an α helix within the bilayer, with hydrogen bonds along the helix axis (inside) and side chains (hydrophobic) pointing into the lipid,

3 hydrophilic stretches (which may be as short as 4–5 amino acids) outside the membrane connect the helices,

4 association between transmembrane helices is brought about by polar interactions, in particular by salt bridges. Positively charged side chains, in the lipid, pair with negative side chains, giving the molecule its tertiary structure.

Points **2** and **4** contrast with the structure of globular proteins, which have charged residues on the *surface* and fold together by associating *hydrophobic regions*. Membrane proteins are sometimes termed 'inside-out' as compared to globular proteins.

These four points have been the basis for modelling the structure of numerous other membrane proteins—in particular for the postulation, in their structure, of particular membrane spanning helices. In reading this text, note that in the case of most other membrane proteins (e.g. the Na⁺ K⁺ ATPase), the proposed structures as yet remain models, while in the case of bacteriorhodopsin, the structure is known.

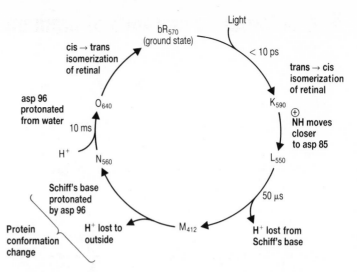

Fig. 40.3. Intermediates in the bacteriorhodopsin photocycle detected spectroscopically. Note the large drop in λ_{max} on deprotonation of the Schiff's base (purple → yellow). Note also that the protein conformational change is not detected spectroscopically.

41 Alternative methods of gradient generation: 2 Primary sodium pumps

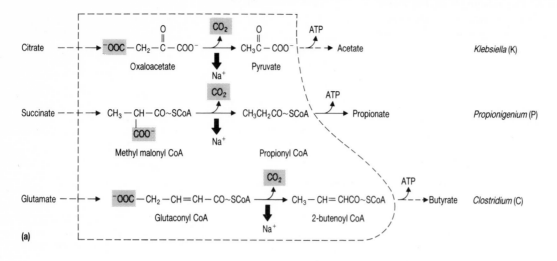

(a)

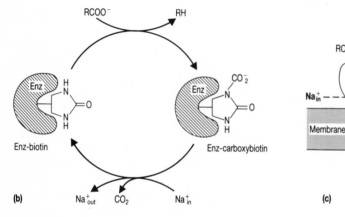

(b)

(c)

Fig. 41.1. (a) Fermentations involving Na^+-linked decarboxylations.
(b) Participation of biotin in enzyme mechanism.
(c) Subunit structure of Na^+-pumping decarboxylase.

H⁺ and Na⁺ gradients

It has been noted several times in this book that a *gradient of any ion* can serve as a source of energy. The expression

$$\Delta\tilde{\mu}_{X^+} = RT \ln c_1/c_2 + F\Delta\psi \qquad \text{(eqn 41.1)}$$

shows this quantitatively; energy stored per mol of (singly charged) ions ($\Delta\tilde{\mu}_{X^+}$) is purely a function of the concentration gradient (c_1/c_2) and membrane potential ($\Delta\psi$), irrespective of their chemical nature.

We have also noted parallels between the use of the most common ions, H^+ and Na^+. In general, bacteria use H^+ gradients to drive transport across the plasma membrane (Section 38) while eukaryotes use Na^+ (Section 42). However, in unusual environments (e.g. high external $[Na^+]$, low external $[H^+]$) the use of H^+ ions for energy *transfer* may lead to energy losses. Some bacterial species which exploit such environments (**marine** bacteria, **halophiles**, **alkalophiles**) have adapted by using a *Na^+-based energy transfer system*. (Note that *not all such bacteria* adopt this strategy—see below.) Bacteria utilizing a Na^+ gradient as energy transfer intermediate therefore require catabolic processes to pump out Na^+, rather than H^+.

Na⁺-pumping decarboxylases

One novel mechanism employed by such organisms to pump Na^+ is a **membrane-bound carboxylase**. Decarboxylations of organic acids (typically β-decarboxylations of oxo acids) are good sources of energy, yielding about 40 kJ/mol *acid* under ambient CO_2 concentrations. Note, however, that this is an efficient use of substrate only under anaerobic conditions; the organisms concerned are carrying out *fermentations* rather than oxidations.

The three decarboxylations so far identified operating as Na^+ pumps are **oxaloacetate to pyruvate** in the fermentation of citrate (e.g. *Klebsiella pneumoniae*), **methylmalonyl** CoA to **propionyl** CoA in the fermentation of succinate (*Propionigenium modestum*), and **glutaconyl** CoA to **butenoyl** CoA in the fermentation of glutamate (*Clostridium symbiosum*) (Fig. 41.1a). In all cases, further energy is obtained from the pathway by using the (high energy) acyl CoA in ATP synthesis, a process more typical of fermentations in general (Section 8).

The mechanism of one such decarboxylation is given in Fig. 41.1b. In the process, CO_2 is transferred to the coenzyme, **biotin**, and it is the loss of this CO_2 which is linked to Na^+ movement.* *Two Na^+ ions* are pumped out per decarboxylation; this could yield up to 20 kJ/mol ions, a value very similar to that of, say, protons in a mitochondrial H^+ gradient. However, measurements indicate that only 11–12 kJ/mol is commonly attained.

The decarboxylases lie on the *inside* of the bacterial cell membrane. The catalytic region is in a globular protein (α subunit, mol. wt. 60 kDa), attached non-covalently to the transmembrane subunits β and γ, which provide the ion channel (Fig. 41.1c). This arrangement is somewhat reminiscent of the $F_1 F_o$ ATPase structure (Section 27).

---

* Note that, in animals, biotin is commonly used by enzymes as a CO_2 carrier in metabolic *carboxylations*. ATP is needed for carboxylation of biotin in these cases—giving a rationale for the use of biotin-CO_2 as a high energy intermediate in ion pumping.

Na⁺-pumping electron transfer

Another source of energy for proton pumping is electron transfer. While protons are the species most commonly pumped by electron transfer complexes (Section 7), there are a small number of oxidative systems known that pump Na⁺. A Na⁺-pumping NADH-Q oxidoreductase occurs in the marine bacterium *Vibrio alginolyticus* (and its close relatives) for example. Probably those electron transfer complexes which operate by a pump mechanism (Section 25) can be adopted for Na⁺ transfer; clearly Na⁺ translocation via a 'loop' mechanism (Section 24) would be impossible.

Uses of Na⁺ gradient by Na⁺-pumping bacteria
(Fig. 41.2)

1 Metabolite transport. The most common use of the Na⁺ gradient in these fermentors is to import the fermentable substrate. For example, citrate enters in symport with Na⁺ in citrate-fermenting bacteria.

2 Motion. Na⁺ may drive the flagellar motor (e.g. in *Vibrio* spp).

3 ATP synthesis. *Propionogenium modestum* possesses a Na⁺-ATP synthase. This is similar to the F_1F_o ATPase of mitochondria and bacteria (Section 27). It has an extrinsic F_1 moiety ($\alpha_3\beta_3\gamma\delta\epsilon$) sensitive to azide, which is similar (the β subunit is 70 per cent homologous) to that of *E. coli*, and an integral membrane F_o (a_2bc_{10-12}), which contains a *DCCD-sensitive Na⁺ channel*.

4 H⁺ extrusion. Na⁺ ions entering through the *Na⁺/H⁺ exchanger* can be used to drive the export of H⁺ ions. This mechanism is used in non-alkalophilic ('neutrophilic') bacteria to regulate cytoplasmic pH; its importance is not clear in the alkalophiles which may need to *import* H⁺ (see below).

Energy conservation in alkalophilic bacteria

Bacteria that live in highly alkaline conditions (pH > 9.0)—the *alkalophiles*—maintain their intracellular pH *lower* than the outside pH, the opposite to most bacteria. This has serious implications for energy conservation. In particular, the two components of $\Delta\tilde{\mu}_{H^+}$, the pH gradient and the membrane potential are *in opposition to each other* (Fig. 41.3a), whereas they are normally additive (Section 5). With the internal pH at 8.5, and the external pH at 10.5, even with a membrane potential of 170 mV (inside negative), the energy released per H⁺ moved inwards is (from equation 41.1):

$$\Delta\tilde{\mu}_{H^+} = RT \ln 10^{-2} - F(0.17) = -5 \text{ kJ/mol H}^+$$

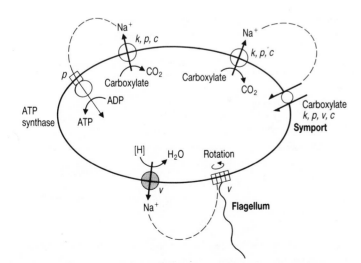

Fig. 41.2. Uses of Na⁺ coupled systems in alkalophiles. [H] → H_2O represents oxidation-linked pump. *K, P, C* and *V* indicate the species concerned. *V = Vibrio alginolyticus*. See Fig. 41.1 for others.

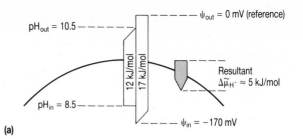

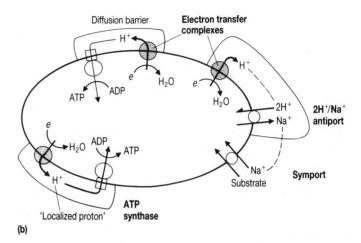

Fig. 41.3. (a) ΔpH and Δψ oppose each other in alkalophilic bacteria. (b) Possible use of H⁺-coupled systems in alkalophiles, e.g. *Bacillus alcalophilus*.

As in other cells, however, alkalophiles need > 55 kJ/mol to synthesize ATP.

To survive in these conditions, the alkalophilic bacteria must adjust their strategies of energy conservation. One possible strategy might be to increase the number of protons employed per ATP synthesized—the H⁺/ATP ratio (Section 21)—to 11 or above. However, this approach is not used in nature; the stoichiometry of H⁺/ATP = 3 seems to be inherent to the mechanism of the ATP synthase. A second strategy is to change the coupling ion; this is used by some alkalophiles, in the Na⁺-dependent systems described above.

However, this strategy is surprisingly not adopted by the extreme alkalophiles (e.g. *Bacillus alcalophilus, Bacillus firmus RAB*), which live at pH 10–11. These organisms have electron transfer chains that pump H⁺ ions outwards, and H⁺-dependent ATP synthases, apparently like those of neutrophilic bacteria.

In fact, the mechanism used for ATP synthesis in these organisms remains obscure. One possibility is that protons yielded in the electron transfer reactions are not released into free solution, but *retained in a localized 'compartment'* which does not equilibrate with the outside medium (Fig. 41.3b). H⁺ in this compartment can be considered to be at high local concentration and, on moving through the synthase, will make ATP in the normal way. The nature of such a compartment is as yet unknown.

Such 'localized protons' would be able to drive ATP synthesis, but, since they must not equilibrate with the medium, *they could not drive transport* via symports or antiports. Here, a change of coupling ion is employed, as in the other alkalophiles. The key component here is a *H⁺/Na⁺ antiporter*, in the membrane of these organisms, *which carries 2 H⁺ inwards in exchange for 1 Na⁺ outwards*. Thus a significant Na⁺ gradient can be established and used to drive transport using Na⁺ symporters and antiporters, as in Fig. 41.3b.

42 ATP-driven ion pumps: an overview

Energy sources for ion pumps

A gradient of ions across a membrane (for example a H^+ gradient across the mitochondrial or bacterial cell membrane), can be used to drive transport across that particular membrane. Using *symport* and *antiport* systems, a wide variety of cations, anions (including organic anions) and neutral molecules (such as sugars) can be moved in this way (Sections 31, 38).

Nonetheless, membranes also use ATP to drive some transport processes. ATP can drive sugar transport in bacteria in addition to the proton gradient, and, indeed, the gradient achieved will be higher (Section 38). More particularly, *ATP is used to move cations* (H^+, Na^+, K^+ and Ca^{2+}) across both prokaryotic and eukaryotic membranes. The enzymes involved are **ion motive ATPases**. They can be classified into *three structural families*.

Classification of pumps

The variety of ion motive ATPases found in animal cells is represented in Fig. 42.1. Superficially, they all look similar; in electron micrographs each appears as a large extramembrane globule with a transmembrane anchor (see Section 26). However, with respect to *mechanism*, they can be divided into two classes; those in which the enzyme becomes covalently *phosphorylated* during catalysis (**P-type**) and those in which enzyme–substrate interactions are strictly *non-covalent* (non-P-type).

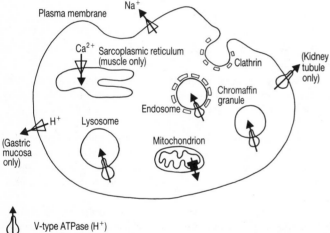

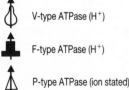

V-type ATPase (H^+)

F-type ATPase (H^+)

P-type ATPase (ion stated)

Fig. 42.1. ATP driven cation pumps in animal cells. The ATP binding site is at the wider end of each symbol.

This mechanistic difference is reflected in a difference in basic *structure* (see below). The P-type ATPases comprise a single, integral membrane protein (mol. wt. $\approx 100\,kDa$) responsible for both ATP hydrolysis and ion movement. In contrast, the non-P-type ATPases have eight or more different subunits, with their ATP binding site on a polypeptide separate from those involved in the ion channel. The P-type enzymes seem more versatile, pumping H^+, Na^+, K^+ or Ca^{2+} while the non-P-type are almost exclusively restricted to H^+.

A convenient experimental criterion for classification among these ATPases is *inhibitor sensitivity*. Only the P-type ATPases are inhibited by micromolar amounts of *vanadate*—a transition state analogue of E–P breakdown. The non-P-type ATPases are inhibited by micromolar amounts of *dicyclohexylcarbodiimide* (DCCD), which blocks their proton channel. However, within the group of non-P-type ATPases, two classes can be distinguished—those inhibited by *azide*, and those inhibited by (micromolar amounts of) *bafilomycin or N-ethyl maleimide* (NEM). The azide sensitive ATPases are termed the **F-type ATPases** (from the F_1F_o ATPases, which fall into this class), and the NEM sensitive are the **V-type ATPases** (the *vesicular* or *vacuolar* ATPases).

The location of these classes of ATPase in animal cells is indicated in Fig. 42.1. P-type ATPases occur on the *plasma membrane* and the *sarcoplasmic reticulum*, while F-type ATPases are restricted to *mitochondria*. V-type ATPases are found typically on the surface of *intracellular vesicles*, pumping H^+ into the vesicles; however in some tissues they are also found on the plasma membrane.

This classification of ion-motive ATPases is not restricted to animal cells but extends over all living organisms. A range of examples (but by no means an exhaustive list) is shown in Table 42.1.

Structure of V-type ATPases

The F-type and P-type ion-motive ATPases are dealt with elsewhere in this text (Sections 27, 43). The structure of the *V-type ATPases* is less well understood. Like F-type ATPases they can be split into a soluble section (V_1) and an intrinsic membrane sector (V_o). In negatively-stained electron micrographs, the V_1 moiety appears as a *ball and stalk* structure on the outside of vacuolar membranes, rather like the appearance of F_1 in mitochondria (Section 27) except that V_1 appears larger (about 12-nm across, as opposed to 9 nm) and has a distinctly grooved appearance (Fig. 42.2a).

The subunit composition of V_1V_o appears to be A_3B_3CDE (abc$_6$) organized probably as in Fig. 42.2b. This again is reminiscent of the F_1F_o ATPase. The *A* subunit (the largest, mol. wt. 70 kDa) is *responsible for ATP binding* and is highly conserved between species. It also contains the reactive cysteine residue that is responsible for this class of ATPase being particularly sensitive to the –SH reagent, NEM (above). Interestingly, the A subunit and the B subunit show some homology with each other (30%), in the same way that the α and β subunits of F_1 are related. The *c subunit* of V_0 (mol. wt. 16 kDa) has a DCCD reactive, intramembrane –COOH group, and

Table 42.1. Distribution of ion-motive ATPases

	F-type ATPases (H^+–ATP synthases)	V-type ATPases (H^+ pumps)	P-type ATPases (ion pumps)
Plants	Chloroplast thylakoid membrane	Tonoplast membrane	Plasma membrane H^+ pump
Animals	Mitochondrial inner membrane	Storage granule Lysosome Endocytotic vesicle Renal tubule plasma membrane	Plasma membrane Na^+/K^+ pump Gastric H^+ pump Sarcoplasmic reticulum Ca^{2+} pump
Fungi	Mitochondrial inner membrane	Vacuole	Plasma membrane H^+ pump
Bacteria	Plasma membrane in *Eubacteria*	Plasma membrane in *Archaebacteria* (V-like ATPase in ATP synthesis)	

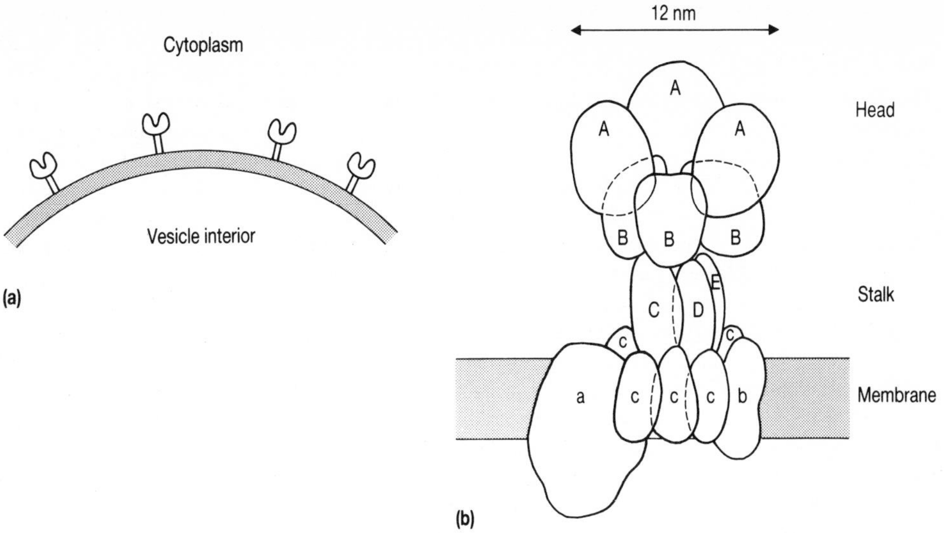

Fig. 42.2. (a) Ball and stalk appearance of V-type ATPase in negatively stained electron micrographs of vesicle.

(b) Schematic subunit organization of V-type ATPase. The number and sizes of the 'stalk' peptides, C, D and E, are uncertain. Subunit a has a mol. wt. of about 100 kDa. Six copies of subunit c are present.

Fig. 42.3. (a) Role of H^+ pumping in accumulation of adrenaline in chromaffin granules. Adrenaline indicated by RNH_2.

(b) H^+ pumping by kidney tubule cells. Passive anion channels indicated by □.

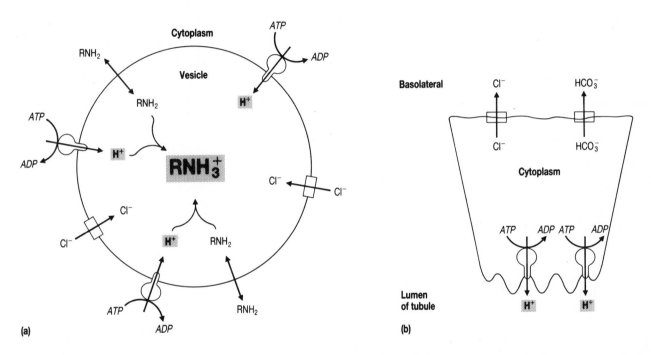

is involved in proton translocation (like the c subunit of F_o). However, unlike F_1F_o, the precise composition as regards the other subunits is not yet certain. The number and/or type of these may vary between species as, for example, when different regulatory peptides are required.

The number of H^+ ions pumped per ATP hydrolysed (H^+/ATP ratio) for the V-type ATPases is not known with certainty, and may vary between species (or tissues). Values of 2 or 3 seem likely.

Unlike F-type ATPases, V-type ATPases have different roles in different tissues—in animals they may pump protons out of cells in the kidney tubules, into chromaffin granules in the adrenals or into lysosomes in phagocytes (see Fig. 41.1). This is associated with different isoforms of the subunits (notably subunits B and E). In other words, although all V type ATPases within one organism have identical A and C subunits, slightly different forms of the other subunits may be incorporated to produce tissue specific isoforms of this enzyme.

Functions of V-type ATPases

The V-type ATPases pump H^+ **into** cell vesicles such as plant cell vacuoles (the *tonoplast* membrane ATPase), fungal vacuoles, and chromaffin granules. In these cases, acidification of the interior provides energy for metabolite accumulation. For example, in chromaffin granules, the uncharged amine, adrenaline, diffuses across the membrane and is trapped, protonated, inside. Members of the same ATPase family are used to switch on acid hydrolases within lysosomes, and to dissociate ligand–receptor complexes in endosomes (for receptor cycling). Note that, for these purposes—in contrast to the uses of P-type and F-type pumps—a pH gradient (*not* $\Delta\psi$) is essential, and thus an anion channel must be present in the same membrane to maximize ΔpH and minimize $\Delta\psi$ (Fig. 42.3a).

An interesting organization is seen in the α (H^+-secreting) cells of the *kidney tubules*. Here the V-type ATPase faces the *lumen*, pumping H^+ ions into it. Anion movement (HCO_3^-, Cl^-) necessary to maintain neutrality occurs at the *opposite pole* of the cell (basolateral side); thus a net movement of H^+ across the epithelium can occur without incurring anion loss from the body (Fig. 42.3b).

Comparisons of pump structure

The *P-type ATPases* (Section 43) comprise a single polypeptide, traversing the membrane 10 times. There is limited (15–20 per cent) sequence homology between class members, but the phosphorylation site, and, to some extent, the ATP binding region, is conserved. There is *no kinase-like motif for ATP binding*. The molecular weight of the active species may be $\approx 200\,kDa$ (dimer of $2 \times 100\,kDa$). Their role *in vivo* is to maintain ion gradients for driving *symports/antiports*, or for *signalling*.

In contrast, both F-type and V-type ATPases have a *bipartite* structure (F_1F_o or V_1V_o) in which the ATP-binding site lies on the extrinsic sector and the ion channel in the transmembrane polypeptides. They show some structural similarities; both F_1 and V_1 contain about five different polypeptides, and F_o and V_o 3 or 4. The overall functional unit, in each case, has a molecular weight of about $500\,kDa$. The catalytic site of each contains a *kinase-like motif* (gly-X-X-gly-X-gly-lys-ser/thr) for ATP binding.

More detailed comparisons, however, show that, although the F-type and V-type ATPases probably *diverged from a single ancestor very early in evolution*, they now form two distinct families. Comparison between F-type ATPases from different species shows the catalytic subunit, β, has a molecular weight of 50–55 kDa, and shows 70% conservation of sequence amongst all species. The V-type ATPases again show a highly conserved catalytic subunit, subunit A, but this has molecular weight 70 kDa; it is distantly related to the F_1–β subunit, but has an *unrelated 150 amino acid insertion* in the centre. Another point of comparison lies in the proton channel; F_o contains a *two helix H^+ conductor* subunit (mol. wt. 8 kDa) while V_o contains a polypeptide of equivalent function (also with one critical –COOH side chain) but twice as long with *four transmembrane helices*. Other subunits, like the γ, δ and ε of F_1 show no homology with the smaller subunits of V-type ATPases.

Finally, there remains a difference in function. The role of the F-type ATPases (of mitochondria, chloroplasts and eubacteria) is to *synthesize* ATP in aerobic respiration—they are pumps working in reverse. The V-type ATPases function as H^+ pumps, they acidify cell compartments. Intriguingly, members of the highly primitive kingdom of **Archaebacteria** use a *V-like ATPase* for their ATP synthesis; in this case however, although the V_1 section resembles the eukaryotic V-type ATPases, the proton channel subunit c is only two transmembrane helices long—as found in the F-type ATP synthases.

43 P-type ATPases: 1 Structural aspects

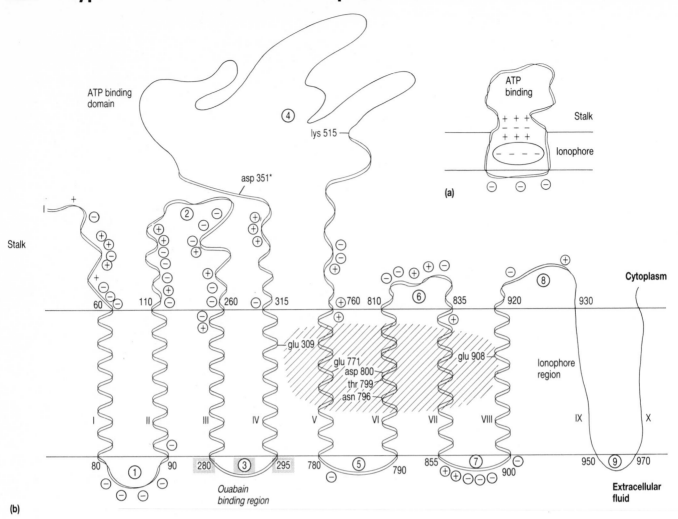

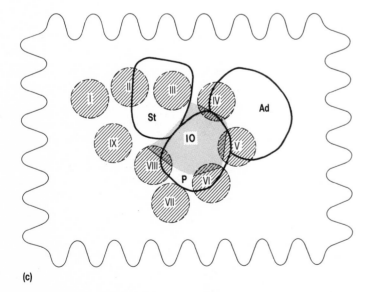

Fig. 43.1. (a) Functional domains of P-type Ca^{2+} ATPases.

(b) Orientation of amino acid chain in P-type ATPases (numbering as for rabbit muscle sarcoplasmic reticulum Ca^{2+} ATPase). The hatched area shows the ionophore region (within the membrane).

(c) Schematic top view of P-type ATPase showing three functional domains *above* the helices grouped around an ion channel. IO = ionophore, P = phosphate binding domain, St = stalk, Ad = adenosine binding domain.

The sodium/potassium and calcium pumps

The two enzymes primarily responsible for maintaining the ionic composition of the cytoplasm, in animals, are the plasma membrane **Na^+/K^+ pump** and the **Ca^{2+} pump** of the sarcoplasmic reticulum. Both use ATP directly as an energy source for ion movements—they are *ion motive ATPases*.

Each enzyme contains a large polypeptide (*c.* 1000 amino acids), which *spans the membrane*, and bears both the *ATP hydrolytic site* and the *ionophore* (the ion transporting region). A smaller, glycosylated peptide is also associated with the Na^+/K^+ ATPase, but has no counterpart in the Ca^{2+} ATPase and is of unknown function.

Both of these ATPases are *integral membrane proteins* and can be isolated only after disrupting the membrane with strong *detergents* (e.g. dodecyl sulphate, $CH_3(CH_2)_{11}SO_3^-$). A good source for the Na^+/K^+ ATPase is (squid giant) nerve axons, or the erythrocyte membrane; the sarcoplasmic reticulum of mammalian muscle is rich in the Ca^{2+} ATPase. The large polypeptides of both are *homologous* and assignment of structural features, below, draws from data obtained on both systems.

Membrane orientation

The *transmembrane regions* of both the Na^+/K^+ and the Ca^{2+} ATPases were established by *sequencing cDNA* (the DNA complement of the mRNA) and identifying hydrophobic regions from *hydropathy plots*. There are *ten transmembrane α-helical segments* (I–X), each about 20 amino acids long (only nine are shown in Fig. 43.1b). The remaining 800–850 amino acids lie outside the membrane, forming loops 1–9.

The *orientation* of the protein was established as follows. Loop 4 is accessible from the *cytoplasm*—it is phosphorylated by cytoplasmic ATP, and can be cleaved by trypsin only in inverted membrane vesicles. Loop 3, in contrast, is labelled from the *outside* of the cell in the case of the Na^+/K^+ pump, by a reactive derivative of the cardiac glycoside, ouabain. Further experiments using monoclonal antibodies against the peptides of the loops, established the arrangement shown in Fig. 43.1.

ATP binding site

(In the following discussion, residue numbers relate to the rabbit muscle sarcoplasmic reticulum Ca^{2+} ATPase.)

The ATP binding site is located in *loop 4*, on the cytoplasmic side of the membrane. This loop is relatively large, containing about half of all the extramembrane amino acids of the enzyme. Near the N-terminal end of this loop is **aspartate-351**, which is *phosphorylated* during the catalytic cycle to the *acyl phosphate*,

$$\text{asp}\overset{\overset{\displaystyle O}{\|}}{-C}-O-\overset{\overset{\displaystyle O}{\|}}{\underset{\underset{\displaystyle O^-}{|}}{P}}-O^-$$

This lies in a sequence which is highly conserved in all P-type ATPases (ser-asp*-lys-thr-gly-thr), asp* being the site of phosphorylation. The *adenosine binding site* lies at the opposite end of loop 4; the probe *fluorescein isothiocyanate* (which competes with ATP for its binding site) covalently labels lys 515. It can be seen therefore, from Fig. 43.1b, that the bulk of the ATPase protein lies on the cytoplasmic side of the membrane. The loops on the opposite side—which face the exterior of the cell in the case of the Na^+/K^+ ATPase, and the interior of the sarcoplasmic reticulum in the Ca^{2+} ATPase—are relatively short.

Perhaps surprisingly, the ATP binding site of the P-type ATPases does not contain motifs like those of other ATP binding enzymes like the F-type and V-type ATPases and the kinases (see Section 42).

Ionophore

The ion binding sites of these pumps are less well defined. In the case of the Ca^{2+} ATPase, the side chains involved are probably *carboxyl groups*, which are known to be *good ligands for Ca^{2+}* in other proteins (e.g. calmodulin, troponin C etc.). Since the ligand requirements for Na^+ and K^+ are more complex, their binding sites cannot be identified from the primary structure.

A large number of carboxyl groups occur in the Ca^{2+} ATPase. Identification of important residues employed site directed mutagenesis of possible residues, expression of the modified protein in cultured kidney cell membranes, and assessment of its transport activity. The conclusions were:

1 Carboxyl groups in extramembrane loops ('stalk region')—particularly loops 2 and 3—have no effect on ion transport.
2 Replacement of carboxyl groups (and some other hydrophilic groups) in the membrane phase (helices IV, V, VI and VIII) leads to complete loss of Ca^{2+} transport. The modified protein in all these cases was shown not to be structurally damaged as it could still be phosphorylated. Only the effects of Ca^{2+} (e.g. in stimulating ATPase activity) appeared to have been lost.

Thus, important groups of the Ca^{2+} ionophore appear to be (some or all of) glu 309, glu 771, asn 796, thr 799, asp 800 and glu 908 (Fig. 43.1). It is possible, by analogy, that the binding sites for Na^+ or K^+ in the Na^+/K^+ ATPases lie in this region, but this is not yet proven.

Three-dimensional organization

The picture given in Fig. 43.1 is a two-dimensional representation of the ATPase protein. We are uncertain as to how the *helices pack* in relation to each other. For example, they may be clustered sequentially, in a horseshoe-type arrangement as in bacteriorhodopsin (Section 40).

Similarly, we do not know *how many* individual polypeptides pack together, within the membrane, to form *one functional pumping unit*. Theoretical considerations, and studies on pump kinetics, suggest that a *dimer* of polypeptides may occur. This is supported by the technique of *radiation inactivation analysis*. In this, the membranes are subjected to high-energy γ-irradiation such that inactivation rate is directly proportional to molecular weight. This indicates a molecular weight of around 250 kDa (i.e. a dimer) for the pump.

However, *reconstitution* studies suggest that the pump may be monomeric. Here, the ATPase protein is reconstituted into liposomes at increasing lipid dilutions. ATPase activity and pumping are observed even when the (random) probability of two ATPase polypeptides being present in one vesicle is very low. This indicates that a dimer is not necessarily the active state of the pump. Further work is obviously necessary to clarify this matter.

44 P-type ATPases: 2 Energetic aspects

Two conformational states of the P-type ATPases

*(The **details** given in this section refer specifically to the Na^+/K^+ ATPase, but its **principles** are generally applicable to any P-type ATPase.)*

Any ion pump must pick up ions on one side of the membrane, and deposit them on the other. There must be two ion binding sites, one exposed to one side of the membrane and one to the other. However, the ion must *bind initially to only one of these sites and be subsequently transferred* to the other. This means that some reorientation of the protein is necessary during the catalytic cycle.

It is unlikely that significant numbers of amino acid residues move in and out of the hydrophobic membrane phase during reorientation. More likely, the transmembrane helices simply twist and/or tilt, accompanied by some tertiary structure changes in the hydrophilic regions—*the reorientating pore model* (Fig. 44.1).

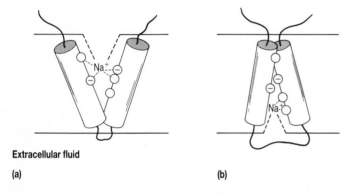

Cytoplasm

Extracellular fluid

(a)   (b)

Fig. 44.1. Reorientating pore model for Na^+ transport.
(a) High-affinity Na^+ binding from cytoplasm.
(b) Low-affinity Na^+ binding on outside.

Structural and kinetic investigations have identified these *two conformational states* of the Na^+/K^+ ATPase, *the E_1 form stabilized by Na^+ ions, and E_2 form, stabilized by K^+*. Structural probes include tryptic digestion (whose pattern changes between the two forms), the fluorescence of a fluorescent label at the ATP binding site (which varies between the two forms; Section 43) and circular dichroism. All three methods indicate significant changes in tertiary structure between forms E_1 and E_2.

Two phosphorylation states of the P-type ATPases

During catalytic turnover, a single aspartate residue (asp 351) is phosphorylated in the P-type ATPases (Section 43). However, two levels of reactivity of the acyl phosphate can be distinguished. $E\sim P$ is a typical acyl phosphate, thermodynamically unstable ('high energy') and capable of reversibly transferring its phosphate to ADP, making ATP ('ADP/ATP exchange'). E-P, however, is atypical—the acyl phosphate is of low phosphate transfer potential (stable) and can be formed spontaneously from E and P_i in solution ('P_i/H_2O exchange').

Studies on the ion sensitivities of these two phosphorylation states in the Na^+/K^+ ATPase indicate that the $E\sim P$ form is stabilized by Na^+ ions—and so corresponds to the E_1 conformation. Similarly, the E-P form, stabilized by K^+, corresponds to E_2. Since the covalent structures of $E_1\sim P$ and E_2-P are identical, stabilization must be achieved by increased *non-covalent bonding* interactions in the E_2-P form. Note also that *ion binding*, and **not phosphorylation**, triggers the change in conformation.

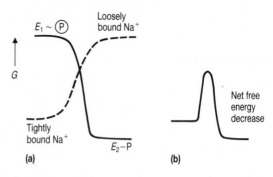

Fig. 44.2. (a) Free energy changes for E-P and Na^+ ions during ion transfer (shown separately).
(b) Sum of changes in a coupled system.

Energy transduction

Bearing these observations in mind, the mechanism of energy transduction (essentially the transfer of free energy from ATP to the ion in question) can be understood in terms of the relationship *between binding affinities and binding energies*. To pick up, say, Na^+ from a region of low concentration (strictly low chemical potential—Section 15), the affinity of the enzyme for Na^+ must be high; $\Delta G^{0'}$ for binding must be very negative. Conversely, to release Na^+, in a region of high concentration, the affinity for Na^+ must be considerably reduced; $\Delta G^{0'}$ for binding Na^+ must be made less negative. Thus *energy is needed to decrease the affinity of the enzyme for Na^+* (Fig. 44.2a). This energy must be derived from the interactions between enzyme, ATP and phosphate.

We have seen, above, that Na^+ ions stabilize E_1 i.e. the *tight binding form for Na^+ is E_1. $E_1\sim P$* contains a typical, high energy, acyl phosphate bond. In E_2-P, by contrast, the acyl phosphate bond is stabilized; it makes additional non-covalent interactions with the protein. In other words, chemical energy in an acyl phosphate bond ($E_1\sim P$) is released to non-covalent bonding interactions in E_2-P (Fig. 44.2a). This energy replaces that lost when the affinity of the enzyme for Na^+ decreases i.e. it is 'used' to decrease Na^+ affinity.

In summary, energy transduction is achieved by energy release (improving the binding of covalently bound phosphate in passing from $E_1\sim P$ to E_2-P) being used to drive ion release (weakening Na^+ binding in passing from $Na^+.E_1$ to $Na^+.E_2$) (Fig. 44.2).

Coupling

An ion motive ATPase must not only be able to hydrolyse ATP and provide an ionophore for ion movement; it must carry out both processes simultaneously so that the energy stored in ATP (chemical energy) is *transduced* into energy stored in the ion gradient. Furthermore, it must hydrolyse ATP *only* when it translocates ions—otherwise both processes could run downhill and energy would be degraded to heat.

For example, the transfer of Na^+ across the membrane, by the mechanism outlined, will be energetically favourable (ΔG slightly negative) (Fig. 44.2b). However, the conversion of $E_1\sim P$ to E_2-P, without transferring Na^+ would be even more favourable (Fig. 44.2a), but physiologically unproductive. This latter process must be prevented, by '*coupling*' the chemical and the transport processes together.

Coupling requirements are met by imposing a *strict sequence of structural changes* on the enzyme during its turnover. This is

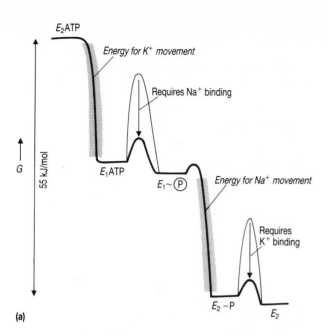

(a)

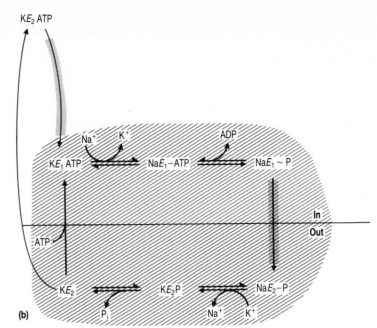

(b)

Fig. 44.3. (a) Energy changes in phosphorylated intermediates for Na$^+$/ K$^+$ ATPase (cf. Fig. 44.2a), showing kinetic limitation (coupling) at ion binding steps.

(b) Classical kinetic scheme for Na$^+$/K$^+$ ATPase (shaded), showing correlation with energetic model.

achieved by *kinetic restrictions (or 'rules'), limiting its behaviour*. In this case, phosphorylation of the enzyme is forbidden without Na$^+$ binding (*Na$^+$ triggers E$_1$ ~ P formation*), and the dephosphorylation is forbidden without K$^+$ binding (*K$^+$ triggers E$_2$-P breakdown*). This is shown by these ions decreasing activation energy for the respective steps (Fig. 44.3a). The resultant sequence is summarized in the familiar kinetic scheme for the Na$^+$/K$^+$ ATPase (Fig. 44.3b). These simple rules do not allow any downhill step ($E_1 \sim P \to E_2$-P, E_2Na $\to E_1$Na etc.) without the corresponding uphill step, and thus couple the system together. Essentially they imply *allosteric activation* of steps in the catalytic pathway by the relevant cation.

Direct and indirect coupling

The transduction mechanism discussed here is an **indirect** one— there is no contact between ATP and the translocated ion, and *energy is transferred via changes in binding interactions within the protein*. It is also a **synchronous** mechanism— *energy is 'lost' from $E_1 \sim P$ and 'gained' by Na$^+$ simultaneously*, avoiding the formation of a metastable 'high-energy conformation' of the enzyme. Both

these ideas fit in with our present ideas of protein structure.

More importantly, analogous coupling mechanisms can operate in a variety of pump systems. For example, the modulation of ATP binding energy during proton pumping by the F_1F_o ATPases (Section 28), and modulation of electron affinity during proton pumping by cytochrome oxidase (Section 25) are directly analogous to this scheme. While direct coupling mechanisms—in which ions and phosphates interact directly—have been proposed, they are less attractive since widely disparate mechanisms would be required for each ion transferred (H$^+$, Na$^+$, Ca^{2+} have widely differing affinities for ATP, for example). In contrast, the Na$^+$/K$^+$ and Ca^{2+} pumps appear very similar in structure and mechanism.

This treatment emphasizes energy transfer between ATP and ions; it implicitly requires that *no energy is required to reorient the protein*. Thus in the energetics, we consider only the binding affinities of E_1 and E_2, ignoring the fact that they face in opposite directions. This is probably reasonable, since any energy needed to change the protein itself would lead to inefficiency in pumping—but it should be remembered that the protein may not behave as a totally frictionless machine in its catalytic cycle.

45 Evolution of bioenergetic systems

Evolution before living organisms—speculations

After formation of the Earth, some 4500 million years ago, life developed. Early on, the basic organic *molecules of life*—acetate, amino acids, phosphate anhydrides and esters and eventually **self-replicating** RNA—became available in the aqueous 'primordial soup'. How they arose is uncertain; the slightly reducing atmosphere would favour the formation of complex molecules, but the occurrence of free water would be deleterious. It may even be that these molecules arose in the dry environment of space, and were carried down to Earth on meteorites.

The chemical reactions of life could not proceed fast enough with these molecules in *dilute* solution. The next stage, therefore, was the appearance of a semipermeable membrane—**self-assembling** from amphiphilic lipids—to enclose regions of this soup and form a primitive 'cell'—a **protobiont**. This arrangement allowed the development of energy-yielding *metabolic pathways*. However, fermentations etc. occurring in this closed cell *tended to lower the pH*, and slow down the reactions; this was countered by the appearance of a **proton leakage channel**. The conversion of this channel into a device *pumping* protons outwards, by addition of an ATPase enzyme, is described below.

Next came the ability of these cells to trap light energy in **photosynthesis**. This had two effects. First, by providing electrons to reduce CO_2, it freed organisms from their dependence on encountering rare organic molecules in the 'soup'. Secondly, by providing a light-driven proton pump, it allowed the reversal of the ATP-driven pump into an ATP synthase. Both of these had the effect of making life forms more plentiful, and from this time (about 3000 million years ago) fossils of unicellular life forms are observed.

Initially, photosynthesis relied on sulphides as electron donors, and involved one photosystem, as in the green sulphur bacteria (Section 18). Later, two photosystems were coupled together, enabling cyanobacterial-like organisms to use *water as an electron donor*. Thus, 2500 million years ago, they laid the basis for oxidative metabolism, by releasing O_2 into the atmosphere (Section 8).

Evolution of proton pumps

The model above suggests that primordial F/V-type ATPases arose, as proton pumps, very early in evolution. Their subsequent evolution (Fig. 45.1) is deduced from homologies between the various polypeptides (Section 42).

1 *The F- and V-type ATPases arose from a common ancestor.* The first stage in their evolution was probably the development of a leakage channel for protons, of about 8 kDa, which prevented cellular pH falling too far during primitive, anaerobic metabolism.

2 *This proton channel combined with an ATPase* (6 copies of a 50 kDa subunit), which increased its effectiveness by **pumping** out the protons. (This explains homology between the α and β subunits of F_1, and the A and B subunits of the V-type ATPases).

3 Gene duplication and divergence led to the appearance of two variants of the 50 kDa subunit.

4 *The F- and V-type ATPases diverged.* A gene duplication occurred, doubling the length of the small subunit in the V-type ATPases. This may have been accompanied by a change in stochiometry; F-type ATPases pump $3H^+$/ATP while V-type ATPases pump fewer ($2H^+$/ATP?).

5 *The F-type ATPases were retained in the plasma membrane of eubacteria*, and became specialized for ATP synthesis. Thus,

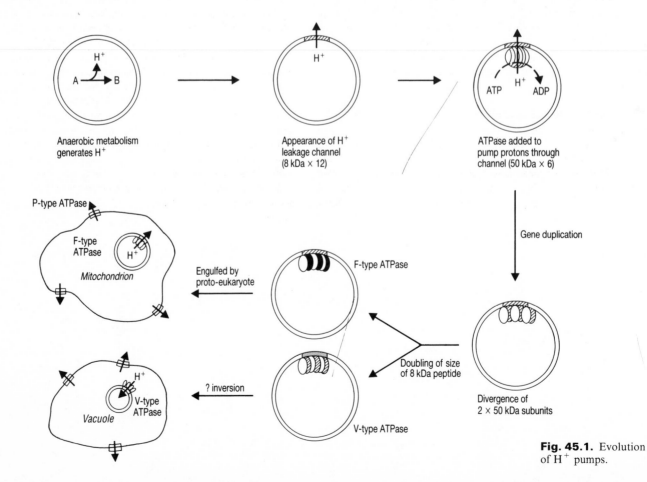

Fig. 45.1. Evolution of H^+ pumps.

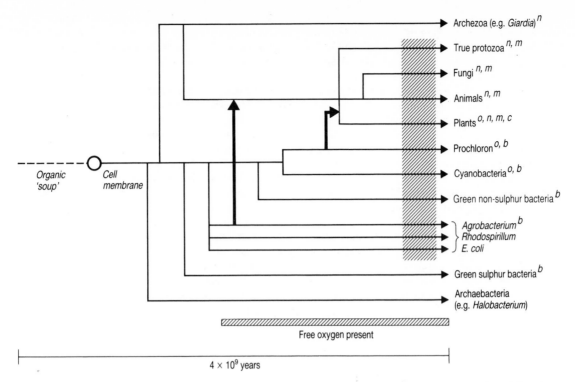

Fig. 45.2. Evolutionary 'tree' from the dawn of life. The thick vertical lines represent endosymbiotic capture of mitochondria (left) and chloroplasts (right). The hatched region indicates organisms adapted to oxidative metabolism. Superscripts indicate possession of: [n] nucleus, cytoskeleton, endomembranes; [m] mitochondria; [c] chloroplasts; [b] member of the Eubacteria.

according to the endosymbiotic theory, below) they occur in *mitochondria and chloroplasts*. The V-type ATPases were retained in the *plasma membranes of archaebacteria*, and transferred (by gene capture?) to intracellular *membranes of vacuoles/vesicles in eukaryotes*.
6 Once inside eukaryotic cells, other subunits (including regulatory subunits) were added to both F- and V-type ATPases over the course of evolution.

Evolution of eukaryotes—origin of mitochondria and chloroplasts (Fig. 45.2)

Primitive, anaerobic cells grew slowly. Their genome was in the form of several pieces of DNA (our present 'chromosomes') each containing functional DNA ('genes') and non-functional DNA ('introns', 'redundant DNA'). Around 3000 million years ago, however, electron transfer was harnessed to provide energy, and fast growing forms, the *bacteria*, appeared. To streamline replication, they eliminated redundant DNA and incorporated the remainder in a single circular chromosome. In fact, two kingdoms of bacteria diverged around this time, the **eubacteria**, and the **archaebacteria**; only in the first, more successful, kingdom did true photosynthesis arise.

The slow growing cells also evolved, developing a true *nucleus* (with several linear chromosomes), and *internal membranes*, such as the endoplasmic reticulum. An ATP-driven Na^+ pump arose (a P-type ATPase), and this took over the maintenance of ionic balance at the cell membrane; the proton pump was partitioned into internal membranes as a V-type ATPase. They also evolved a *cytoskeleton*, allowing them to *endocytose* other organisms; they lived by fermenting this organic material. They became primitive members of the third present day kingdom, the **eukaryotes**.

Due to photosynthesis, oxygen appeared in the atmosphere about 2500 million years ago. To the primitive eukaryotes (*proto-eukaryotes*), this was a severe hazard (see Section 8). Bacteria, however, had electron transfer chains, and had adapted to use oxidations for energy—and were thus able to use oxygen to exploit new habitats where light did not penetrate. *Oxidative electron transfer chains evolved from photosynthetic ones*; hence their similarity (Sections 5, 18).

It can be seen that association between the slow-growing fermenters and the oxidative bacteriae would be advantageous. The proto-eukaryotes needed to engulf the bacteria to obtain organic material, but also, in fermentation, provided end products which the bacteria could oxidize. This somewhat precarious situation was ultimately resolved, about 1500 million years ago, when a proto-eukaryote developed an **endosymbiotic** association with a bacterium. Here, bacteria were engulfed but, rather than digested, fed fermentation products by the eukaryote which, in turn, extracted from them high-energy compounds, such as ATP. During further evolution, these bacteria lost their cell walls and transferred most of their genome to the eukaryotic nucleus; the *obligate endosymbiont*, the **mitochondrion** and the true eukaryote was born around 1000 million years ago.

A similar argument applies to the origin of **chloroplasts**, through endosymbiotic association of bacteria capable of oxygenic photosynthesis with a proto-eukaryote. Chloroplasts are believed to have evolved later than mitochondria (all plant cells have mitochondria too), and have transferred rather less of their DNA to the nucleus.

Nearest neighbours and missing links

In this model, it is proposed that cells once existed in which oxidative or photosynthetic bacteria lived symbiotically inside primitive eukaryotic cells. Interestingly, such associations are still found today. The protozoan giant amoeba, *Pelomyxa palustris*, harbours within its cell membrane aerobic bacteria which distribute when the cell divides; a unicellular endosymbiotic system reminiscent of mitochondria. As regards chloroplasts, an alga, *Cyanophora paradoxa*, contains cytoplasmic bodies very similar to cyanobacteria in structure (even with a cell wall), which carry out its photosynthesis. These *cyanelles* have lost some genes to the nucleus, and it is tempting to see them as equivalent to a stage in chloroplast evolution.

Present day organisms may mimic evolutionary stages, but are unlikely actually to be such stages frozen in time. Chloroplasts, for example, (containing chlorophyll *a* and *b*, and light harvesting complexes, Section 16) are probably not derived from cyanobacteria (containing chlorophyll *a* only, and phycobilisomes). A more likely ancestor may have given rise to the present day family of the free living **prochlorophytes**, such as *Prochloron*, which carry out oxygenic photosynthesis and possess chlorophyll *a* and *b*, and stacked thylakoids.

Mitochondria are likely to have evolved from the α subclass of the **purple (non-sulphur) photosynthetic bacteria**, as deduced from similarities in cytochromes *b* and *c*. rRNA sequencing has indicated that the nearest present day bacteria are exemplified by *Agrobacterium tumifaciens*, which, interestingly, may still enter into symbiotic associations with present day eukaryotes.

What of the other half of this relationship, the primitive proto-eukaryote? Possible candidates occur in the class of primitive protozoa, the **Archezoa**. These organisms lack organelles, and have small, 70S ribosomes, but do possess a nucleus, cytoskeleton, and internal membranes (see above). Examples are the *microsporidia*, for example the dysentery-causing *Giardia*. These organisms may never have needed oxidative metabolism for optimal growth, due to their parasitic lifestyle, and have perhaps remained closer in form to the proto-eukaryote.

Evidence for the endosymbiotic origin of mitochondria and chloroplasts

Mitochondria and chloroplasts show the following characteristics which resemble those of bacteria rather than eukaryotes.

1   Circular DNA chromosome.
2   No cholesterol in membrane.
3   Small (70S) ribosomes lacking 5S RNA.
4   Protein synthesis inhibited by chloramphenicol (antibiotic).
5   Proteins synthesized with formyl methionine as N-terminus.
6   ATP-driven H^+-pump in membrane.
7   Where there is an equivalent macromolecule in the organelle and cytoplasm of eukaryotes (e.g. rRNA, superoxide dismutase, glyceraldehyde phosphate dehydrogenase in chloroplasts), sequencing shows the bacterial protein/RNA to be related to the organellar species rather than that of the eukaryote cytoplasm.

Reading list

General texts

Cramer, W.A., Knaff, D.B. (1990) *Energy Transduction in Biological Membranes.* Springer-Verlag, New York.

Harold, F.M. (1986) *The Vital Force—a Study of Bioenergetics.* W.H. Freeman, New York.

Nicholls, D.G., Ferguson, S.J. (1992) *Bioenergetics 2.* Academic Press, London.

Mitochondrial and bacterial electron transfer

Anraku, Y. (1988) Bacterial electron transfer chains. *Ann. Rev. Biochem.* **57**: 101–113.

Babcock, G.T., Wickstrom, M. (1992) Oxygen activation and the conservation of energy in cell respiration. *Nature* **356**: 301–309.

Cammack, R. (1992) Iron–sulfur clusters in enzymes: themes and variations. *Adv. Inorg. Chem.* **38**: 281–322.

Esposite, M.D., de Vries, S., Crimi, M., Gehlli, A., Patarnello, T., Meyer, A. (1992) Mitochondrial cytochrome b—evolution and structure of the protein. *Biochim. Biophys. Acta* **1143**: 243–271.

Moore, G.F., Pettigrew, G.W., Rogers, N.K. (1986) Factors influencing redox potentials of electron transfer proteins. *Proc. Natl Acad. Sci. USA* **83**: 4998–4999.

Moser, C.C., Keske, J.M., Warncke, K., Faird, R.S., Dutton, P.L. (1992) The nature of biological electron transfer. *Nature* **355**: 796–802.

Schneider, H., Lemasters, J.J., Hochli, M., Hackenbrock, C.R. (1980) Fusion of liposomes with mitochondrial inner membranes. *Proc. Natl Acad. Sci. USA* **77**: 442–446.

Walker, J.E. (1992) The NADH : ubiquinone oxidoreductase (Complex I) of respiratory chains. *Q. Rev. Biophys.* **25**: 253–324.

Wendeloski, J.J., Matthew, J.B., Weber, P.C., Salemme, F.R. (1987) Molecular dynamics of the cytochrome c/b$_5$ electron transfer complex. *Science* **238**: 794–796.

Photosynthetic systems

Allen, J.F. (1992) How does protein phosphorylation regulate photosynthesis? *Trends Biochem. Sci.* 12–17.

Anderson, J.M., Andersson, B. (1982) The architecture of photosynthetic membranes—lateral and transverse organisation. *Trends Biochem. Sci.* 288–292.

Barber, J. Andersson, B. (1992) Too much of a good thing: light can be bad for photosynthesis. *Trends Biochem. Sci.* 12–17.

Deisenhofer, J.J., Michell, H. (1991) High resolution structures of photosynthetic reaction centres. *Ann. Rev. Biophys. Biophys. Chem.* **20**: 247–266.

Golbeck, J.H. (1993) The structure of photosystem I. *Curr. Opinion. Struct. Biol.* **3**: 508–514.

Kuhlbrandt, W., Wang, D.N. (1991) 3-Dimensional structure of plant light harvesting complex determined by electron crystallography. *Nature* **350**: 130–134.

McDermott, G., Prince, S.M., Freer, A.A., Hawthornthwaite-Lawless, A.M., Papiz, M.Z., Cogdell, R.J., Isaacs, N.W. (1995) Crystal structure of an integral membrane light harvesting complex from a photosynthetic bacterium. *Nature* **374**: 517–521.

Nitschke, W., Rutherford, A.W. (1991) Photosynthetic reaction centres, variations on a common structural theme. *Trends Biochem. Sci.* **16**: 241–245.

Rutherford, A.W. (1989) Photosystem II, the water splitting enzyme. *Trends Biochem. Sci.* **14**: 221–232.

Energy transduction and proton pumps

Brand, M.D., Reynafarje, B., Lehninger, A.L. (1976) Stoichiometric relationship between energy dependent proton ejection and electron transport in mitochondria. *Proc. Natl Acad. Sci. USA* **73**: 437–441.

Ferguson, S.J. (1986) The ups and downs of P/O ratios. *Trends Biochem. Sci.* **11**: 351–353.

Thayer, W.S., Hinkle, P.C. (1975) Kinetics of ATP synthesis in bovine heart submitochondrial particles. *J. Biol. Chem.* **250**: 5336–5342.

Trumpower, B.L. (1990) The protonmotive Q cycle—energy transduction by the cytochrome bc$_1$ complex. *J. Biol. Chem.* **265**: 11 409–11 412.

Trumpower, B.L., Gennis, R.B. (1994) Energy tranduction by cytochrome complexes in mitochondrial and bacterial respiration. *Ann. Rev. Biochem.* **63**: 675–716.

Wickstrom, M. (1984) Pumping of protons from the mitochondrial matrix by cytochrome oxidase. *Nature* **308**: 558–560.

F$_1$F$_o$ ATP synthase

Abrahams, J.P., Leslie, A.G.W., Lutter, R., Walker, J.E. (1994) Structure at 2.8 Å resolution of F$_1$-ATPase from bovine heart mitochondria. *Nature* **370**: 621–628.

Boyer, P.D. (1989) A perspective of the binding change mechanism for ATP synthesis. *FASEB J.* **3**: 2164–2178.

Fillingame, R.H. Molecular mechanics of ATP synthesis of F$_1$F$_o$-ATP synthases. In: Krulwich, T.A. (Ed), *The Bacteria*, vol. 12, pp. 345–349. Academic Press, San Diego.

Senior, A.E. (1990) The proton translocating APTase of *E. Coli*. *Ann. Rev. Biophys. Biophys. Chem.* **19**: 7–41.

Integration

Brand, M.D., Murphy, M.P. (1987) Control of electron flux through the respiratory chain in mitochondria and cells. *Biol. Rev.* **62**: 141–193.

Capaldi, R. (1988) Mitochondrial myopathies and respiratory chain proteins. *Trends Biochem. Sci.* **13**: 144–148.

Harris, D.A., Das, A.M. (1991) Control of mitochondrial ATP synthase in heart. *Biochem. J.* 561–573.

Walker, J.E. (1992) The mitochondrial transporter family. *Curr. Opinion. Struct. Biol.* **2**: 519–526.

Wallace, D.C. (1992) Diseases of the mitochondrial DNA. *Ann. Rev. Biochem.* **61**: 1175–1212.

Other ion-motive systems

Dimroth, P. (1987) Sodium ion transport decarboxylases and other aspects of Na ion cycling in bacteria. *Microbiol. Rev.* **51**: 320–340.

de Rosier, D.J. (1992) Whipping flagellin into shape. *Curr. Opinion. Cell. Biol.* **2**: 280–285.

Henderson, P.J.F. (1993) The 12 transmembrane helix transporters. *Curr. Opinion. Cell. Biol.* **5**: 708–721.

Henderson, R., Baldwin, J.M., Ceska, T.A., Zemlin, F., Beckmann, E., Downing, E. (1990) Model for the structure of bacteriorhodopsin based on high resolution electron cryo-microscopy. *J. Mol. Biol.* **213**: 899–929.

MacLennan, D.H. (1990) Molecular tools to elucidate problems in excitation—contraction coupling. *Biophys. J.* **58**: 1355–1365.

MacLennan, D.H., Brandl, C.J., Korczak, B., Green, N.M. (1985) Amino acid sequence of a Ca^{2+}/Mg^{2+} ATPase from rabbit muscle sarcoplasmic reticulum, deduced from its complementary DNA sequence. *Nature* **316**: 696–700.

Nelson, N. (1992) Organellar proton ATPases. *Curr. Opinion. Cell. Biol.* **4**: 654–660.

Pedersen, P.L., Carafoli, E. (1987) Ion-motive ATPases. *Trends Biochem. Sci.* **12**: 146–150, 186–189.

Poolman, B., Konings, W.N. (1993) Secondary solute transport in bacteria. *Biochim. Biophys. Acta* **1183**: 5–39.

Schuster, S.C., Khan, S. (1994) The bacterial flagellar motor. *Ann. Rev. Biophys. Biomol. Struct.* **23**: 509–539.

Schwartz, R.M., Dayhoff, M.O. (1978) Origin of prokaryotes, eukaryotes, mitochondria and chloroplasts. *Science.* **199**: 395–403.

Tanford, C. (1983) Mechanism of free energy coupling in active transport. *Ann. Rev. Biochem.* **52**: 379–409.

Index